HOW TO BUILD

Build a Stable

RED BARN TOOL HOUSE

BY DONALD R. BRANN

Library of Congress Card No. 72-88710

THIRD PRINTING – 1975
REVISED EDITION

Published by
DIRECTIONS SIMPLIFIED, INC.

Division of
EASI-BILD PATTERN CO., INC.
Briarcliff Manor, N.Y. 10510

FIRST PRINTING

REVISED EDITION
1973

ISBN 0-87733-679-2

NOTE
Due to the variance in quality and availability of many materials and products, always follow directions a manufacturer and/or retailer offers. Unless products are used exactly as the manufacturer specifies, its warranty can be voided. While the author mentions certain products by trade name, no endorsement or end use guarantee is implied. In every case the author suggests end uses as specified by the manufacturer prior to publication.

Since manufacturers frequently change ingredients or formula and/or introduce new and improved products, or fail to distribute in certain areas, trade names are mentioned to help the reader zero in on products of comparable quality and end use. The Publisher.

CAN YOU AFFORD A HORSE?

Less than 60 years ago, owning a horse was an important economic necessity. Horses provided transportation, the power to plow and cultivate. As this nation became mechanized, horse ownership became a popular hobby until their initial cost, keep, feed and housing reached a point where only the well-to-do could afford the luxury of their company. Today, by an odd set of circumstances, owning a horse is something many middle income families can again afford, if they are willing and able to provide the care and shelter. Almost everyone who lives in an area where codes permit building a stable, now find stabling facilities in big demand. Not only can a small stable be self-supporting, but also a "life saver".

A family with a growing child, one that's "crazy about horses", has a unique opportunity of guiding the child into a healthy and happy maturity through some of the most difficult years any child ever faced. An interest in horses is a time consuming hobby that helps transform a child into an individual who can stand alone, who no longer needs the recognition of her peers to feel accepted. A child who develops into a strong individual, one who doesn't need to be part of a group, who keeps in constant communication with his family, is truly a blessing.

Don R. Brann

TABLE OF CONTENTS

BLESSED ARE THOSE WITH AN INTEREST

If you like horses, or happen to be parents of a child that loves horses, you already know the special bond that exists. Encourage this interest and you could create some of life's happiest memories. An interest in horses may seem like an expensive hobby, and yet it seldom is. Most youngsters who develop a fondness, willingly work to support their hobby, and the hobby helps create a healthy supporter.

Building a stable not only simplifies boarding a horse, but also shapes the life of the owner, the destiny of the family. My bride's love of horses was the motivating force that catapulted us into building our first stable almost two years before we could afford a horse. Months of spare time was invested clearing a site, digging trenches, putting in footings, foundation, framing, etc. Hours normally consumed by cocktail parties, golf, and other social activities was invested in construction. Long before the stalls were complete, two paying boarders were ready to move in, and the income was helpful during the lean years. Many years later we sold the property and received a handsome capital gains for the man hours of labor and materials invested.

Charting a path through today's jungle of problems, is a task relatively few parents are prepared to cope with. While encouraging a child to develop an interest in horses may seem costly, another family, with the same age child and income, soon discovers solving a drug problem can be far more expensive. Putting money into materials, and materials into a shelter of any kind, is like putting money into a bank that pays interest, plus a Capital Gains.

Building a stable establishes new responsibilities. These not only concern you, your neighbors and the community, but also the horses you board. As soon as the building is completed and in operation, you, and/or your child or children, will begin to live a kind of life only a chosen few are privileged to enjoy. So don't louse it up by being too lazy to keep a stable as spotless as you do your kitchen, office, or person. How you maintain a stable spotlights you as an individual. Show respect for the animals, and people will have a lot more respect for you.

Those who neglect cleaning a stall daily soon discover the urine and droppings soften the earth floor. You then have a smelly mess that attracts a lot of flies, complaints from neighbors, and invites a reprimand from the board of health. If you don't rectify the problem immediately, you could be fined for maintaining a health hazard.

Unless you are thoroughly familiar with the problems of stable operation, and can satisfactorily answer questions a building inspector, or the board of health can raise during a hearing, your permit could be postponed or rejected. The faster you answer questions intelligently, the more convincing will be your application. Read and study this book before filing for a permit.

Building provides hours of satisfying effort and almost complete escape. While many folks will build this barn themselves, with help when needed, others will hire a mason or carpenter to help them get started. By working alongside an experienced man, you not only get the job done faster, but also gain valuable knowledge. If you employ help, you become a contractor. Talk to your insurance agent. Tell him what you plan on doing. Follow his advice concerning temporary liability and compensation insurance. If you live in an apartment, and building a barn is something in your future file, build a balsa wood model to prove you can do it. Working ¾" to 1' scale with balsa provides an interesting way of teaching yourself how to build.

If you have a child that loves horses, build a slightly larger model using 1" to one foot scale. This builds a play stable measuring 20" x 30", perfect for the budding horseman. While it takes lots of room, the child will enjoy endless playtime hours.

Building is fun because it provides healthy relaxation, exercise, a great feeling of accomplishment, while it opens doors to new fields of endeavor. But remember one important piece of advice. When you do anything you are normally unaccustomed to doing, it's important to work only as long as you feel physically fit. Stop as soon as you become tired. It is at this point accidents and errors occur that can be avoided. Always get adequate assistance when you raise wall frames, or lift anything heavy. Don't take a chance and overtax your strength. For safety's sake, keep a First Aid Kit handy at all times. Always use lumber free of knots when building scaffolding. Make certain temporary braces are nailed securely before walking, or working on any raised platform.

OBTAIN A PERMIT

Since a building permit is needed in most areas, obtain one from the town clerk before starting construction. While step-by-step directions cover foundation work and framing, no attempt has been made to provide directions for plumbing or wiring since codes in many areas require this work be done by licensed craftsmen. If you build in a section where no license is required and wish to do this work yourself, read Book #694 Electrical Repairs Simplified and Book #682 How to Install an Extra Bathroom. Both can help you plan this work.

This book explains how to build a three box stall stable that's only 6 feet longer than many two car garages. Any additional ten foot modules can be added to its length. While zoning boards normally issue permits for a garage, a permit for a stable may be refused for any number of reasons. Since you can't build without a permit, and a permit is normally granted to those who know what they are doing, and who abide by the approved regulations, it's essential you read all the preliminary information contained on the following pages. This will help you intelligently answer all questions a zoning committee may ask.

If you live in, or close to a development, your neighbors' opinions get high priority. Before filing for a permit, talk to all neighbors and get their approval. If neighbors' children are invited to participate in the stable activities before you approach the parents, chance for acceptance may overcome initial jealousy. If your neighbors are agreeable, but you are concerned about zoning, get up a petition. Ask everyone to sign. In your request for a building application, explain how you plan on removing manure weekly, and that the building will in no way create a health hazard.

Permission to build is based on many considerations. The building must be sufficient distance from your house, property line, and neighboring houses. It must not be located near a well, adjacent to a stream, or run-off that feeds into a fresh water lake.

The site selected must be in an area where drainage from stalls can be absorbed within your property, and not pollute your neighbors' land. The site must provide convenient access for a hay truck. It must provide outdoor storage for at least a week's load of manure. And the manure must be stored where a heavy rain doesn't drain off onto your neighbor's property, or feed into community water supplies.

After selecting a site that meets zoning restrictions and provides easy access for a truck, position the stable on the site so outside stall doors are on the protected side. Note how the snow banks up against one side of your house and face the least useable side of the stable in this direction.

These are some of the primary problems that arise at most building permit hearings. Strong objections by near neighbors are difficult to overcome. They have just as much, and frequently more legal reasons to object. Some stable owners, like people who picnic, are pigs who have no regard for others. A poorly kept stable can be an eyesore, a health hazard, and can help destroy property values. On the other hand, a stable can add the quality of country living to a suburban area, and thereby greatly increase its desirability.

In many communities, the board of health must also approve permission to build a stable. Long before you file for a building permit, nose around, learn the facts. Talk to the board of health, find out what they require. Then when you file, you will have all the proper answers to any question that can arise.

One of the first board of health requirements concerns drainage from a stable to a dry well, Illus. 1; outdoor storage and removal of manure. If you plan to build on sufficient acreage, you won't face many of these problems. Those who build in a fairly restricted area can do what the author did.

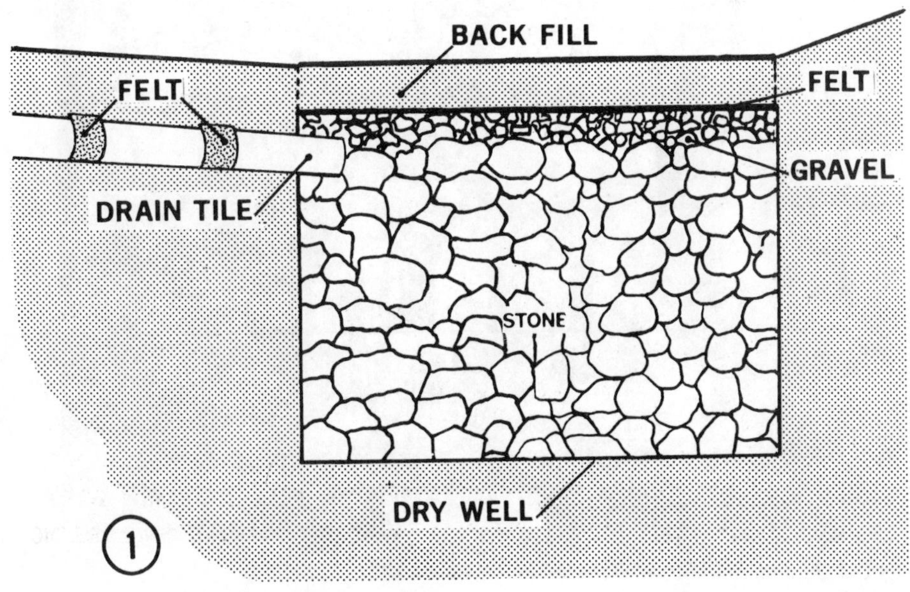

①

Through mail order, he purchased a four wheel chassis, and built a manure wagon, Illus. 2, using 2 x 10 x 12'. The body was built 22½" high. By adding extra sideboards, the wagon accommodated a full week's load of manure from eight horses. After contacting several commercial florists, one finally agreed to make a weekly pickup. Sometime later, a young neighbor decided to grow mushrooms. He got permission to build a lean-to shed and with booklets explaining how to grow mushrooms from the Dept. of Agriculture, he started a business now in its 20th year.

②

Since removal of manure is frequently the key to your getting a permit to build, solve this problem—find out who wants what. Contact landscape gardeners, commercial florists, or encourage a business executive who, because of pressure he can't cope with, to grow mushrooms before he drinks himself to death. When you overcome the initial odor, mushroom growing relieves tension, provides much healthy exercise plus a steady income with practically no pressure.

SITE SELECTION

To meet most zoning restrictions, the site you select must not only be sufficient distance from your property line and house, but also from neighboring houses. It must also meet all restrictions laid down by the board of health. The site must be level, or be leveled, and preferably one that's not at the base of a hill. Good drainage away from your building is important.

If the site selected is low, and holds water after a rain, it won't make a very good place to build. If site slopes, and you can bulldoze an area that's 6" to 12" higher than surrounding land, it could make a perfect site. Always play safe and locate the stable where it's higher than surrounding ground. Build a dirt ramp. Floor level should be at least 6" above grade.

If your stable plans include a heated tack room, or an apartment in the stable, you may want to bury an oil tank adjacent to foundation. A hole for a 500 to 1000 gallon oil tank provides a lot of extra dirt fill.

Be sure to trench for a water line below frost level before laying footings or foundation block. Armored underground electric cable and a telephone extension line, can be buried along with a copper or plastic water line.

If you plan on installing a toilet, sink, or other plumbing fixture, you will need a trench for the waste line to either a sewer connection or septic tank. Since a water line must be buried below frost level, if the site selected is rocky, running a water line may present some problems. While a rocky stable site can be made useable when covered with sufficient dirt fill, a water line must be buried below frost level.

Plumbing supply stores sell pipe insulation that normally provides sufficient protection when covered with a moderate amount of earth. Unless a water line is adequately protected against a freeze-up, don't attempt to lay it.

TOOLS NEEDED

The following tools will be needed. Pick, shovel, hoe, cement mixing tub (make one 3' x 6' x 1' deep), mason's trowel, mason's float, ball of building line, plumb bob, line level, carpenter's level, framing square, miter square, hammer, brace, ⅝", ¾" bits, crosscut and rip saws. Fifty foot steel tape, chalk line, hatchet, folding zig-zag rule, nail set, ¼", ½", ¾", 1" wood chisels. Wrecking bar. Adjustable end wrench, screw driver, two extension clamps. Nailing apron, overalls, 26" and 5' horses; buy adjustable sawhorse clamps and cut legs to length required. Step ladder, plus a 12 or 14 ft. ladder. This can be made by nailing 1 x 4 rungs on two 2 x 3 or 2 x 4's. Space rungs about one foot apart. Nail securely to uprights with 8 penny common nails.

In many areas, building material and tool rental stores rent small concrete mixing machines, electric hammers and pipe scaffolds. These are great time and labor savers and are well worth the rental fee. An electric hand or table saw is also an excellent investment. These are normally equal to an extra man's time.

MAKE PLOT PLAN

Most building departments require a plot plan showing location of proposed building and size. They also want a floor plan and elevations of the front, back, right and left side, Illus. 4, 5, 6, 7, 8. Draw a sketch similar to Illus. 3, when filing for a plan. Fill in distance X and Y. If you want to extend the building, do so in two foot modules.

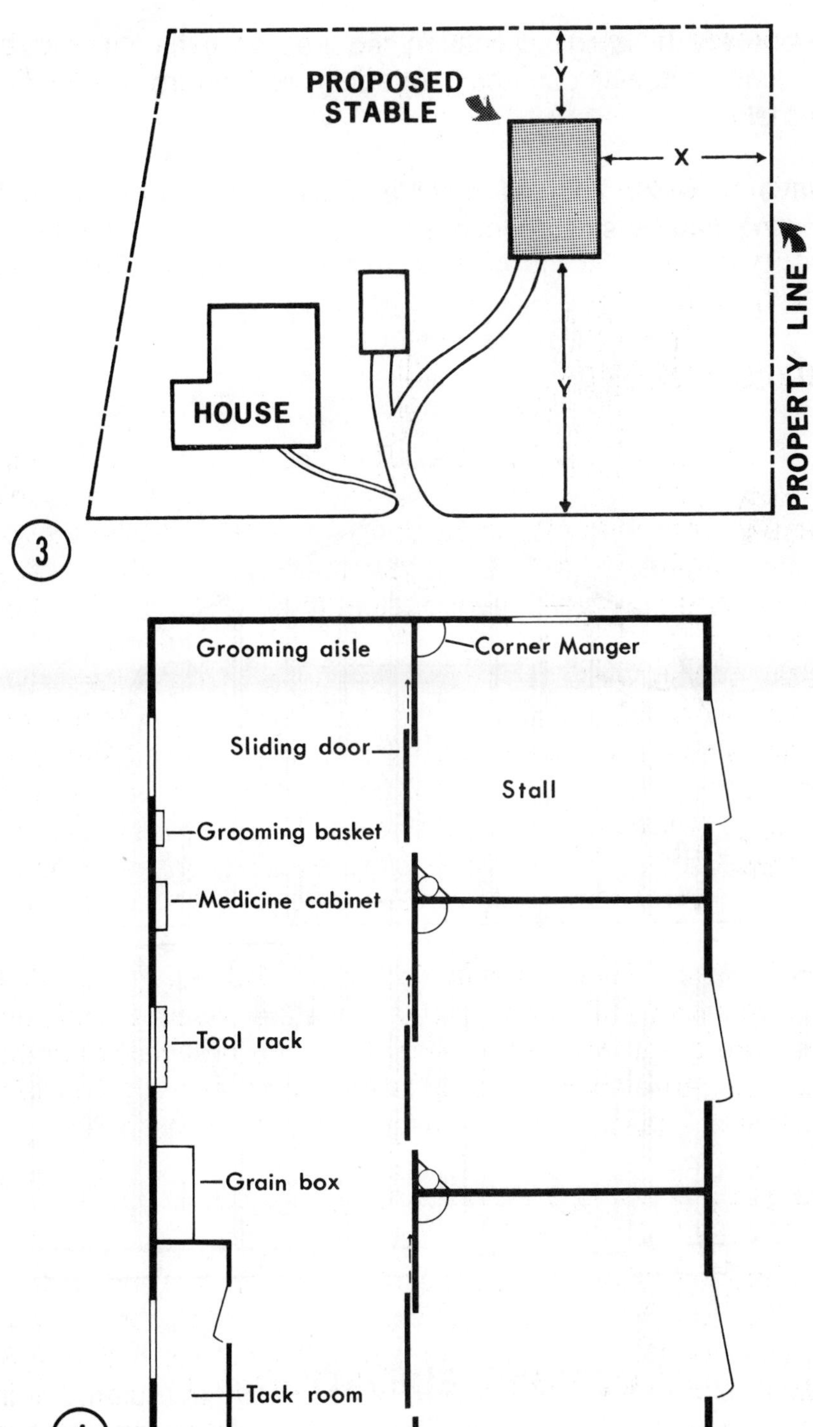
PROPOSED
STABLE
Y
X
Y
PROPERTY LINE
HOUSE
3
Grooming aisle
Corner Manger
Sliding door
Stall
Grooming basket
Medicine cabinet
Tool rack
Grain box
Tack room
4

Since every building application requires an estimate of cost, your lumber dealer can estimate same working from the list of materials.

Always consider yourself outside, facing the building, when following step-by-step directions.

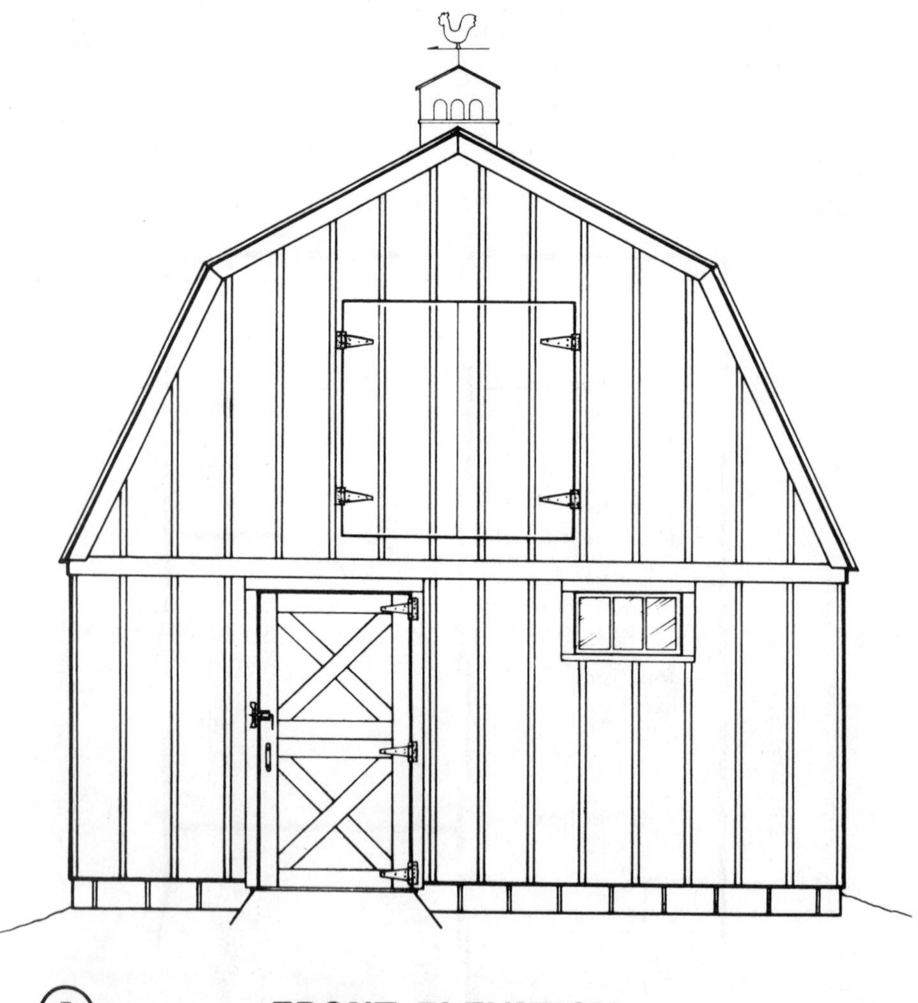

(5) **FRONT ELEVATION**

Due to the variance in lumber width and thickness always check length of inner framing members before cutting. We figured 2 x 4 as measuring 1½" x 3½", 2 x 6 as measuring 1½" x 5½" and 2 x 8 as measuring 1½" x 7¼".

Page 151, contains complete list of materials for 20x30' stable.

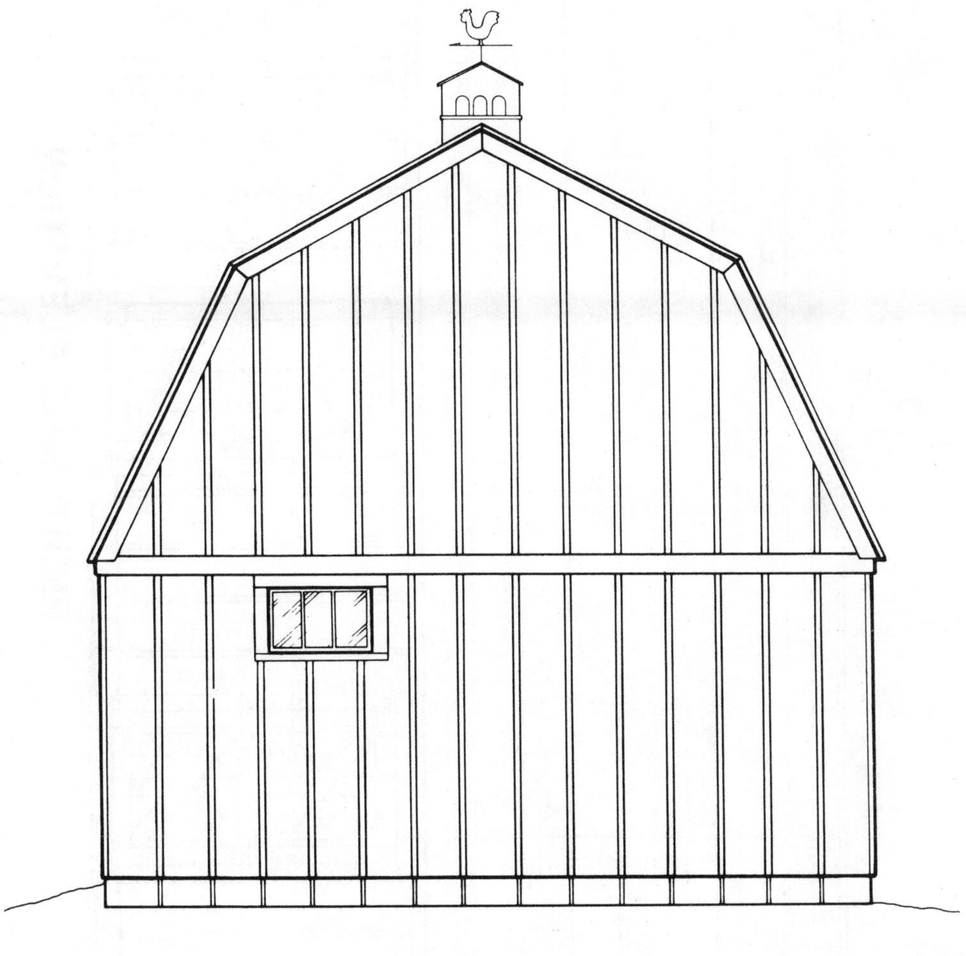

(6) **REAR ELEVATION**

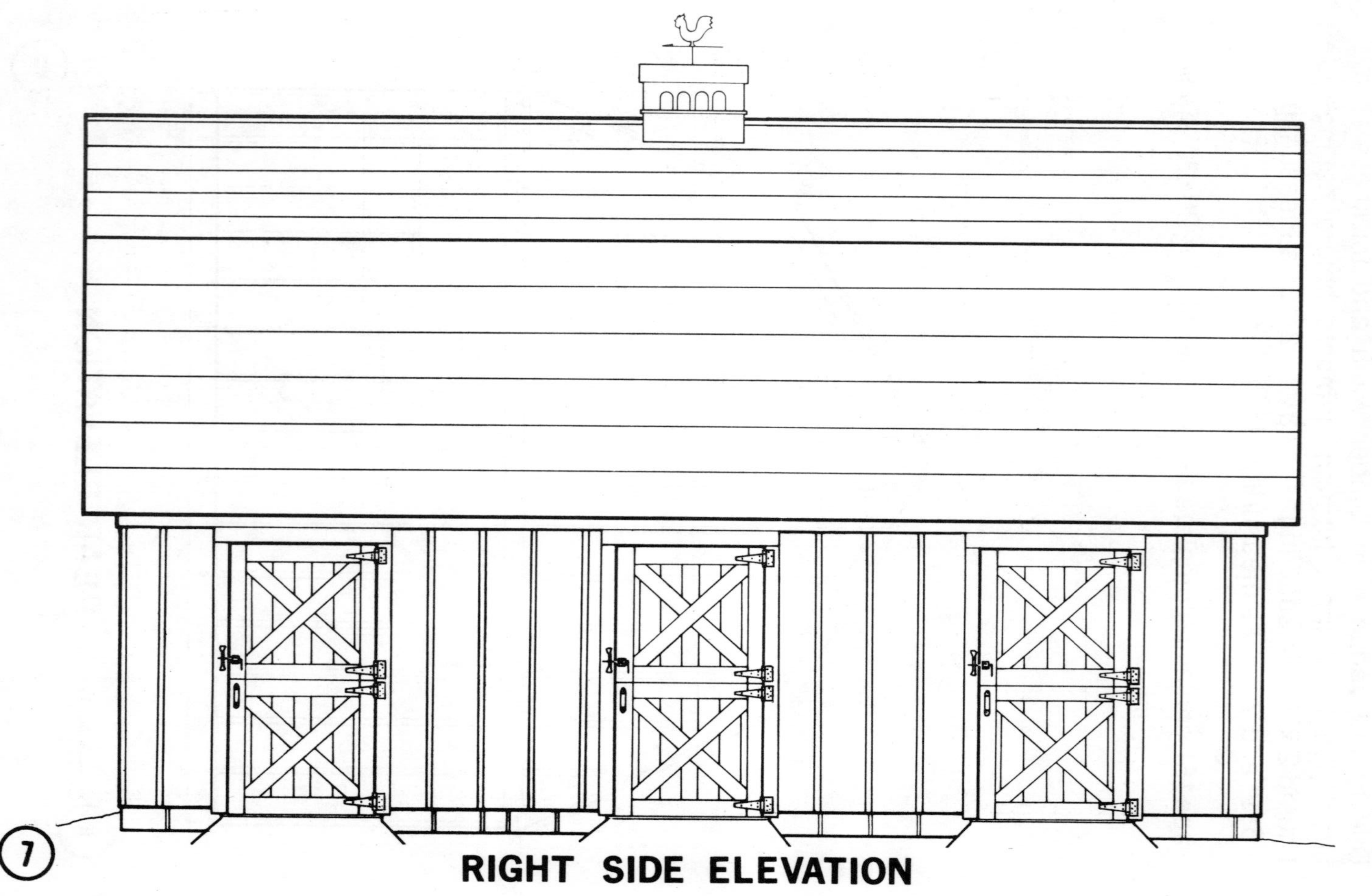

RIGHT SIDE ELEVATION

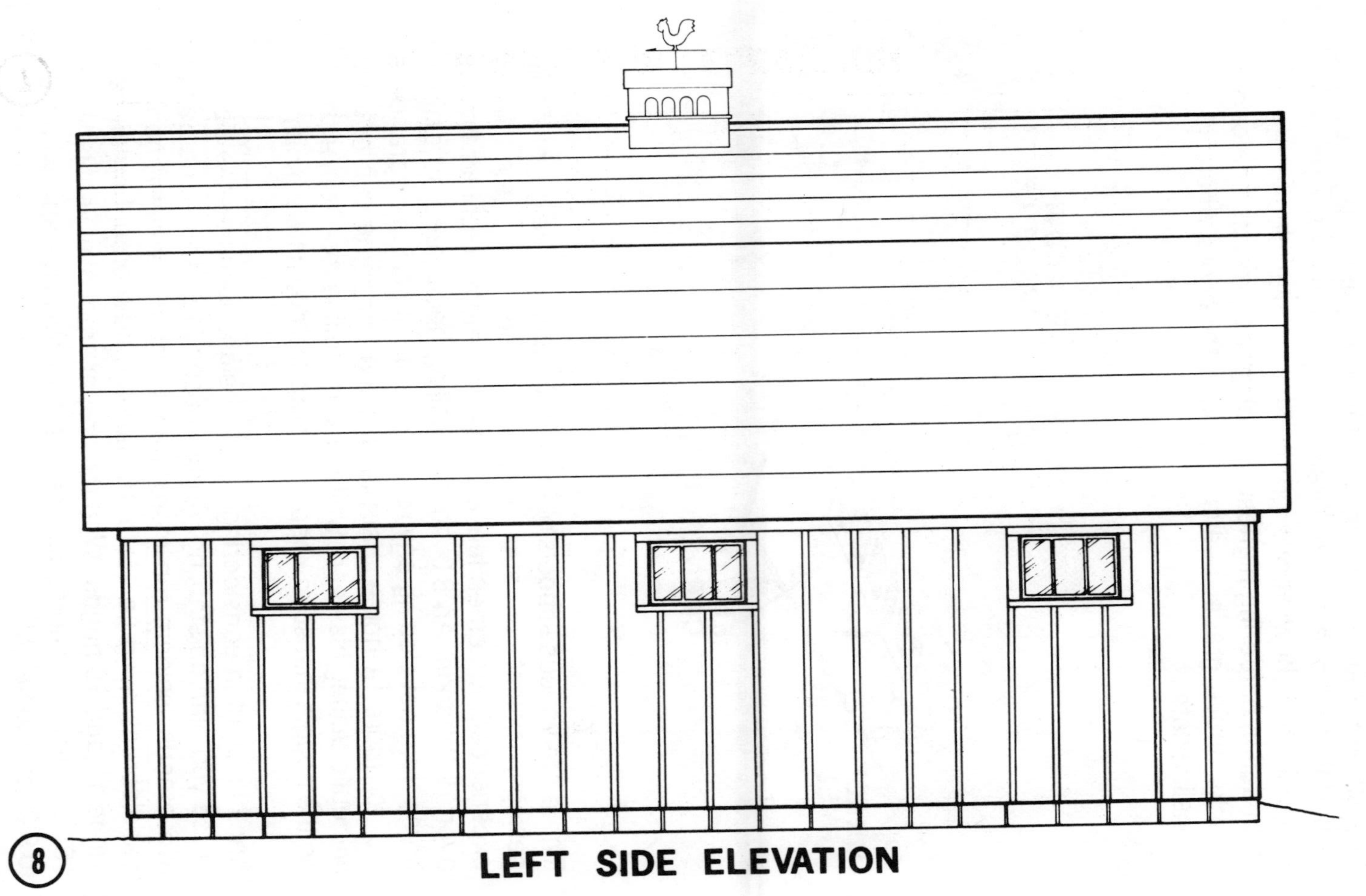

LEFT SIDE ELEVATION

To obtain the headroom a stable requires, lay out your foundation so one, two, or three 8" courses of 6 x 8 x 16" blocks, Illus. 9, 10, show above grade. If your site slopes, be sure foundation bordering highest part of grade projects one or more blocks above grade. Leave openings for doors where illustrations indicate.

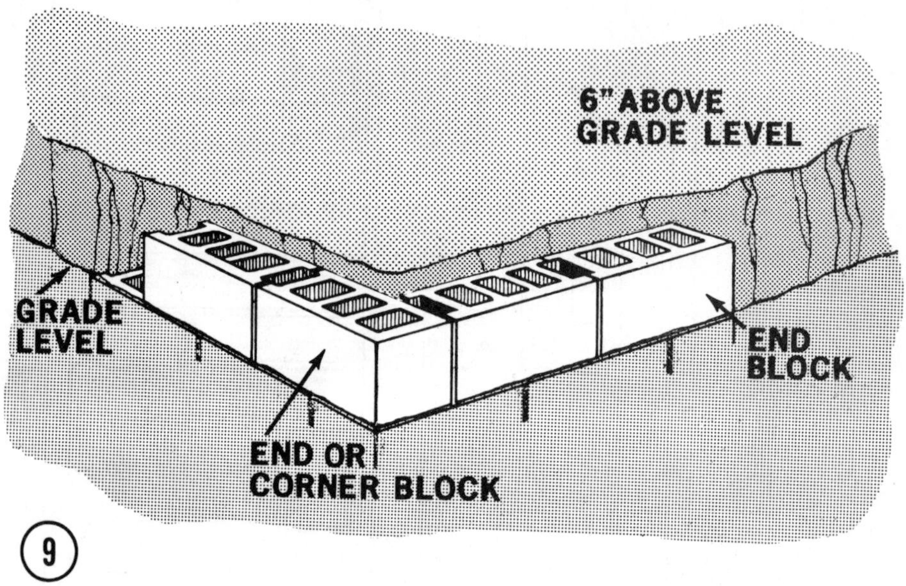

There are various schools of thought concerning height of stable ceilings, height of stall walls, dividers, etc., etc. Many experienced horsemen feel 8'0" is a minimum height for horses over 15.2 hands. An 8'0" to 8'6" ceiling helps keep a fractious horse from developing any rearing habits. If ceiling clearance is 9'0" or more, a horse that plays, soon learns he can rear up without cracking his head; while an 8'0" ceiling results in a knock on the noggin that soon keeps their rearing down to a minimum.

Another point in considering ceiling height concerns the size of horse you plan on stabling. Arabians and Morgans, to name but two breeds, seldom run more than 15 hands, while American Saddlebreds average 15 to 16 hands, and thoroughbreds can run to 17 and 18 hands. With the proper amount of ventilation, a 16 hand and over horse can be comfortable in a stable with 8'6" to 9'0" ceiling height.

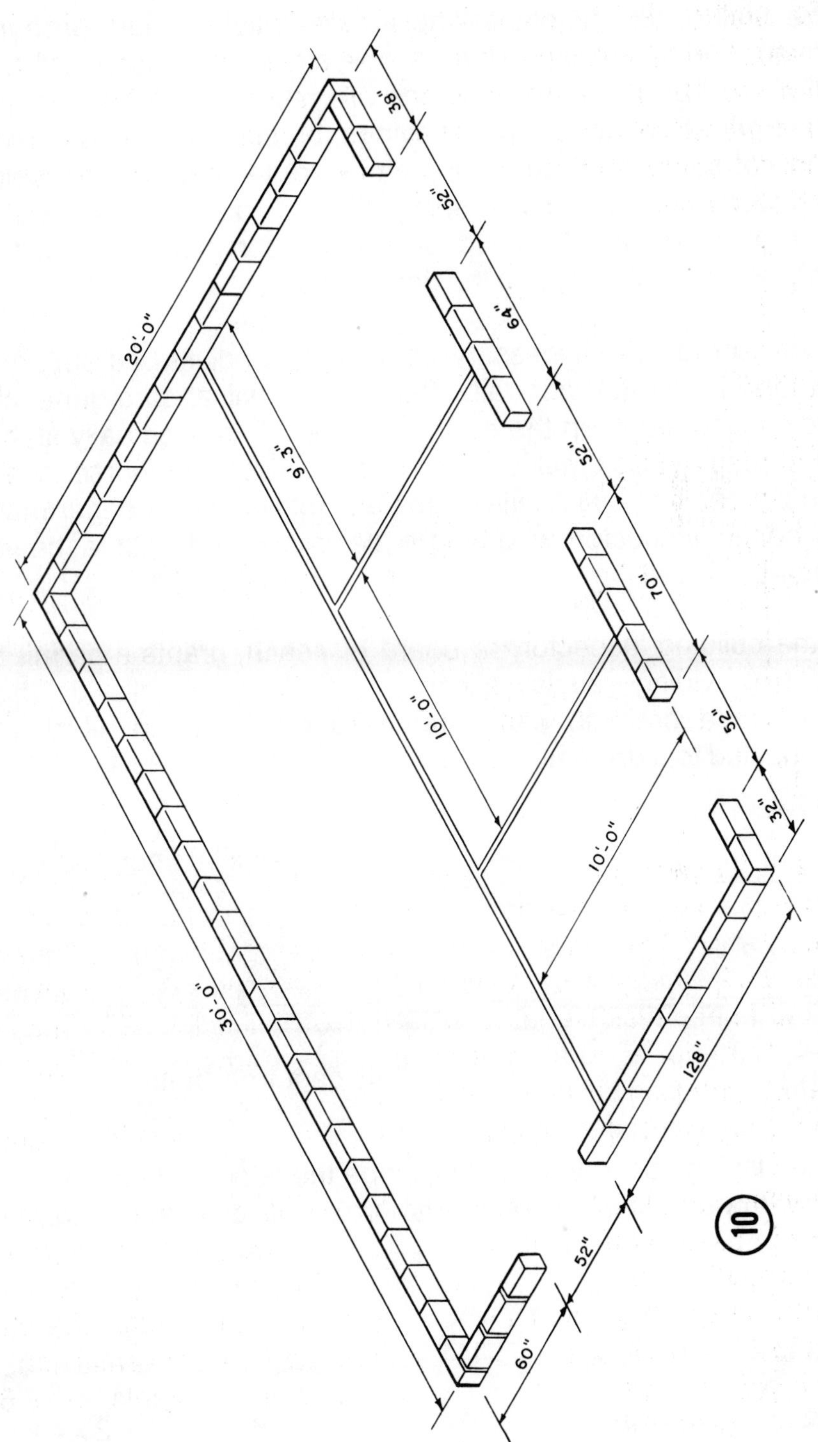
20'-0"
30'-0"
9'-3"
10'-0"
10'-0"
38"
52"
64"
52"
70"
52"
32"
128"
52"
60"
10

The stall floor, the place where horses live, should be earth covered with bedding. This can be straw, if you can find and afford to buy it. Wood shavings are recommended when you can't get straw. If you buy shavings, be sure you get shavings and not sawdust. Wood chips can also be used. But shavings and wood chips should be sprayed lightly to settle dust. Dust is murder on horses so sawdust is not recommended, and can only be used if it's kept on the damp side.

If you spread the necessary bedding on undisturbed soil, and the stall is mucked out daily, the bedding will absorb urine and droppings, and keep the earth floor firm. This once daily stable care keeps stalls smelling "like a rose" insofar as horse lovers are concerned. It also eliminates any friction with a department of health inspector who might be concerned with a health hazard.

If the building inspector, or board of health, grants a permit to build, providing you lay a concrete floor in all stalls, do this. Connect a drain in each stall to a master drain, Illus. 11, connected to a dry well, Illus. 1.

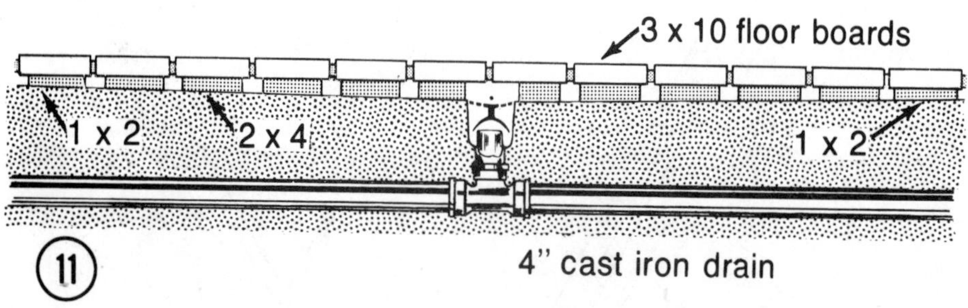

When laying concrete in a stall, slope the floor ⅛" to every foot, either toward a center drain, or to drain outside of stall, Illus. 12.

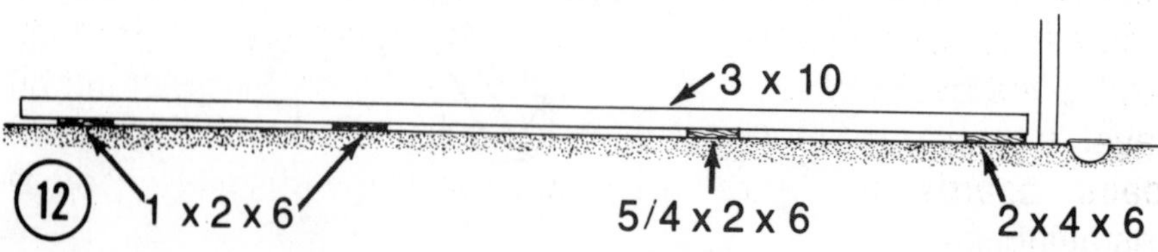

Since concrete is too hard on horses' legs, lay 3 x 10 planks the full width of the stall, Illus. 13.

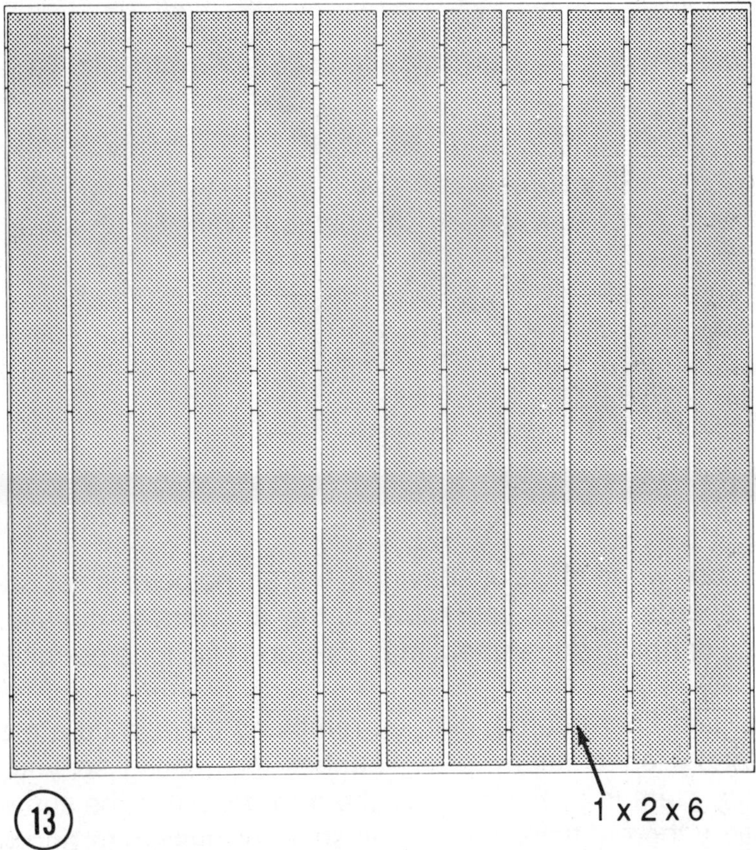

Cut these 2" less than overall length required, then nail 1 x 2 x 6" blocks, Illus. 14, to one edge of each plank. This separates the floor boards from each other. To keep planks off the floor, nail 1 x 2 or 2 x 4 blocks to bottom of floor boards. This supports boards above concrete, allows urine to drain from the bedding and gives bedding a longer lease on life. Floor boards must be level. Use all weather pressure treated lumber for floor boards or paint boards with creosote or wood preservative before installing.

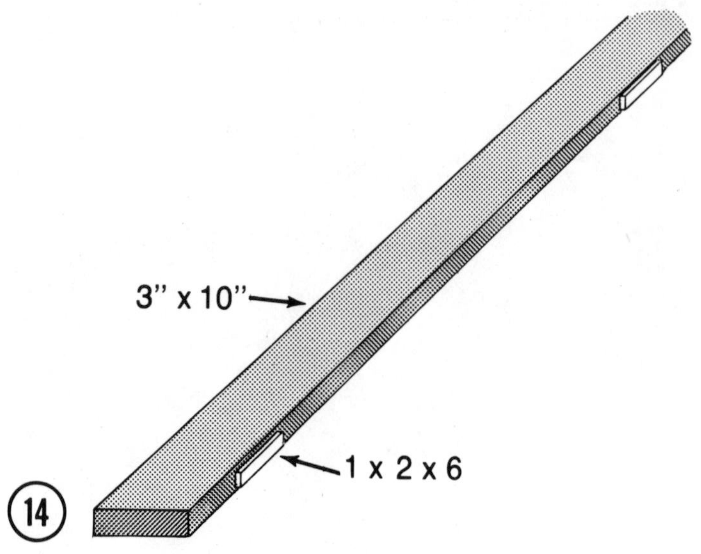

If you use wood chips or shavings, there's no point in separating floor boards. Unless you know all the answers to questions a zoning or health department may ask, don't file for a permit.

If you use floor boards, plan on removing them once a month, preferably on a sunny day. Number the planks so you can replace them in position. Allow them to dry out in the sun while you do a thorough cleaning of the stall. Always check and clean the center drain.

WHAT SIZE STABLE SHOULD YOU BUILD

Many years ago, when horses worked six days every week, dairies and other commercial enterprises stabled horses in straight stalls. These were frequently 52" in width by 7'6" to 9'0", Illus. 15, for a medium weight horse. A straight stall that's too wide encourages a horse to roll. Since they are tied, they can easily get cast, caught in a position they can't extricate themselves, and in some cases choke to death with the tie line.

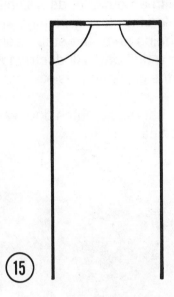

A straight stall is acceptable for riding stables where horses are worked daily, and in those barns adjacent to pastures where a horse can be turned out when he isn't ridden. If you build a private stable and expect to keep the horse indoors some days, and have no adequate pasture to let him run in, build box stalls. These can be as small as 8'0" x 8'0"; but 10'0" x 10'0" is better.

As the floor plan, Illus. 4 indicates, three slightly smaller than 10'0" x 10'0" box stalls can be incorporated into a 20'0" x 30'0" building. This is a practical plan since it provides storage for hay above, a grain box, tackroom, feed area, space for tools and other stable gear. Larger stalls require much more bedding, take longer to muck out and are more costly to maintain. A 12' x 12' or 10' x 14' stall is ideal for a mare and foal. We have had mares foal in 10' x 10' stalls, and experienced no problems. Since the entire bed must be removed, along with the afterbirth effects, a large stall does increase costs. Always select a large stall for a mare due to foal. Never attempt to move a mare and foal to a new stall for at least a week or longer after birth. When they have become accustomed to going to pasture, you can return them to a new stall.

Each stall is planned with an outside, as well as an inside door. The inside sliding door leading to a grooming aisle is a big help. Grooming in the aisle helps keep the stall cleaner. It also gives the horse a change on a dull day. All doctoring, blacksmithing, etc., should be done in the grooming aisle.

Outside doors, Illus. 16, should be dutch doors and hinged. Door construction is shown on page 113.

16

Actual construction follows this general procedure:

1. Obtain a building permit.
2. Lay out guide lines for foundation.
3. Excavate perimeter foundation trenches.
4. Run plastic or copper water line, electric, telephone and drainage lines where required.
5. Pour footings, lay concrete block foundation walls, position anchor bolts or anchor clips.
6. Build forms, pour footings for aisle wall and stall dividers.
7. Build and raise exterior wall frames; erect center aisle beam.
8. Install ceiling joists.
9. Sheath exterior walls.
10. Apply flooring in loft.
11. Prefabricate and raise rafters.
12. Apply sheathing to roof.
13. Apply #15 felt, fascia, roofing.
14. Install window and door frames, windows, doors.
15. Build stalls.
16. Nail plyscord ceiling panels (optional).
17. Install water hydrant, feed mangers, water pails.
18. Lay paving in passageway (optional).
19. Frame in tack room, feed room, grain bin, oat storage room in loft.
20. Have a barn warming, but don't allow smoking.

BEFORE STARTING CONSTRUCTION read through the step-by-step directions completely at least twice. If any step isn't entirely clear, ask your lumber dealer to explain. Once you know how the building is constructed, you will be able to more easily make changes in the floor plan. Study each illustration so you become familiar with all parts and their position in relation to others.

GENERAL DIRECTIONS

To meet building code requirements without a lot of conversation, studs should be spaced 16" on centers, except where doors, windows or stall dividers require alternate placement. The rafter construction, Illus. 72-88, suggests using metal fasteners available at lumber yards. These greatly simplify and strengthen construction.

The design of this barn provides a lot of usable space in the loft. As every stable owner soon learns, the more space you have to store hay and straw, the cheaper is its cost. Construction of the rafter is explained on page 75.

Use care when laying out guide lines for foundation trenches. Trenches should be square, bottom level, and an equal distance from guide lines. Use the layout square to guide you in making right angles at all corners.

To make a layout square, square off ends of two 1" x 6" x 8 ft. boards. Mark exactly 4' and 3' on one, 5' on the other. Nail or screw together as indicated. If you work with precision, you will have an accurate right angle, Illus. 17.

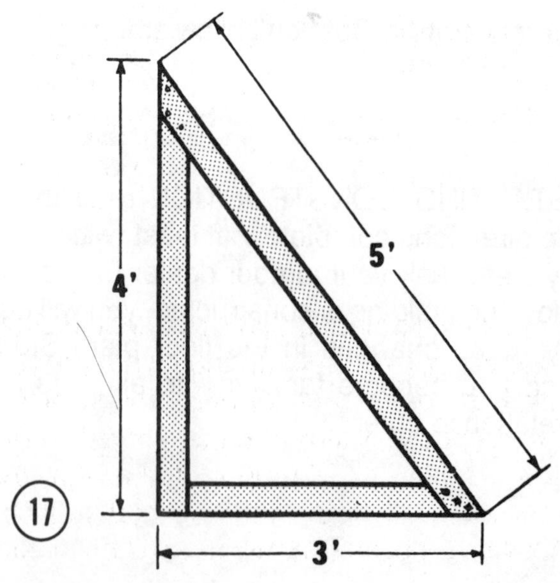

Use a straight 2 x 4 with a carpenter's level to guide you in leveling trenches, Illus. 18.

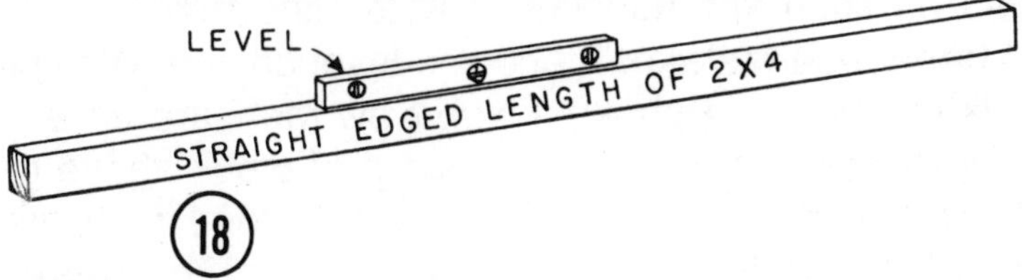

LAYOUT, EXCAVATING

After selecting site, decide what grade level (level of ground) you want to maintain around finished stable. Drive a stake flush into ground at this level. Drive a nail into stake, allow it to project ½" to 1" above stake. This is called a grade level stake, Illus. 19. Tie a line to nail. Place a line level on line.

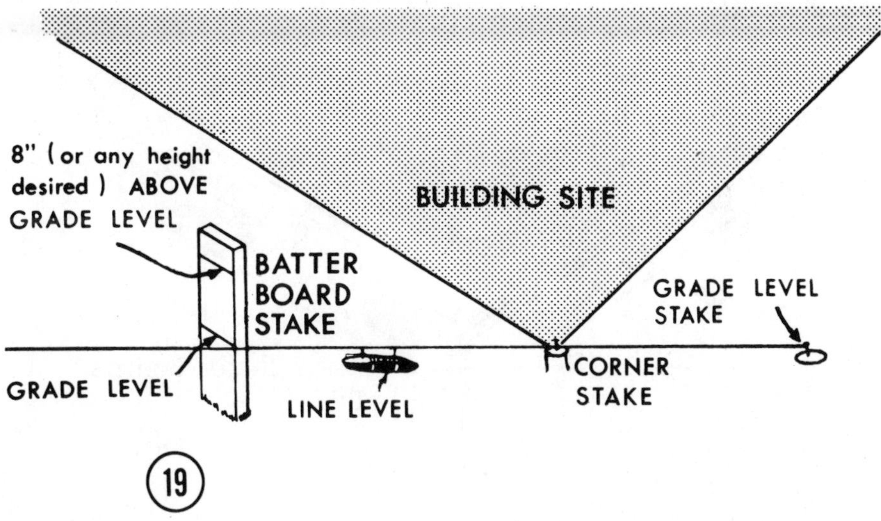

Stretch a line to various points within site selected. Note low and high points. This will give you some idea where to remove soil, where to fill in. Now drive a stake into ground at a point that represents a front corner. This stake can project 3" or 4" above ground. Drive a finishing nail in top of stake. Approximately 3 feet from corner stake, drive three more stakes to hold batter boards. Use 1 x 4 or 1 x 6, sharpened at one end for stakes.

Drive into ground. These stakes should equal or exceed height of proposed foundation. They can project above batter boards or be sawed off flush, as indicated. Stretch line from grade level stake to batter board stake. Place line level on line. With pencil, mark stakes when line shows level. Now measure up 8". This mark now represents top of one course of block above grade, Illus. 10. Nail batter boards to stakes at this point if you want to maintain an 8" height above grade; or at a point that represents two, three or as many 8" courses of block above grade as you need. The top edge of batter board should indicate top edge of foundation.

Place end of steel tape on finishing nail in front corner stake. Measure across twenty feet and drive other front corner stake. Drive finishing nail into top of stake at exact point indicating width of stable. Erect batter boards 3 ft. away. The front building line should be parallel to or at right angles to house or property line, Illus. 20.

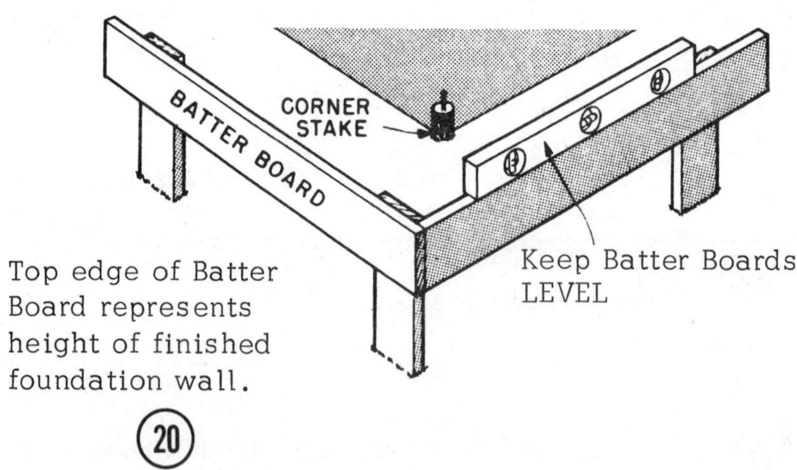

(20)

With a plumb bob to guide you, Illus. 21, stretch a line directly over finishing nails in corner stakes. Hold point of bob over nail in corner stake. Foundation line should touch plumb bob line. Fasten a stone or wrap line around batter board to maintain taut line. This line indicates top outside edge of foundation.

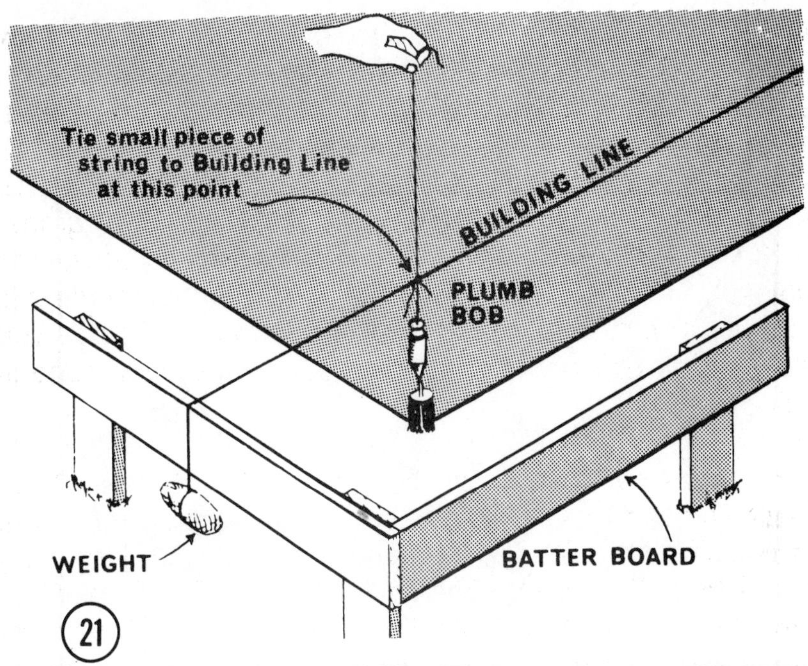

Check line with line level. Saw slot in batter board to lower line; fasten line to a nail in batter board to raise. Follow same procedure to lay out other lines, Illus. 22, 23.

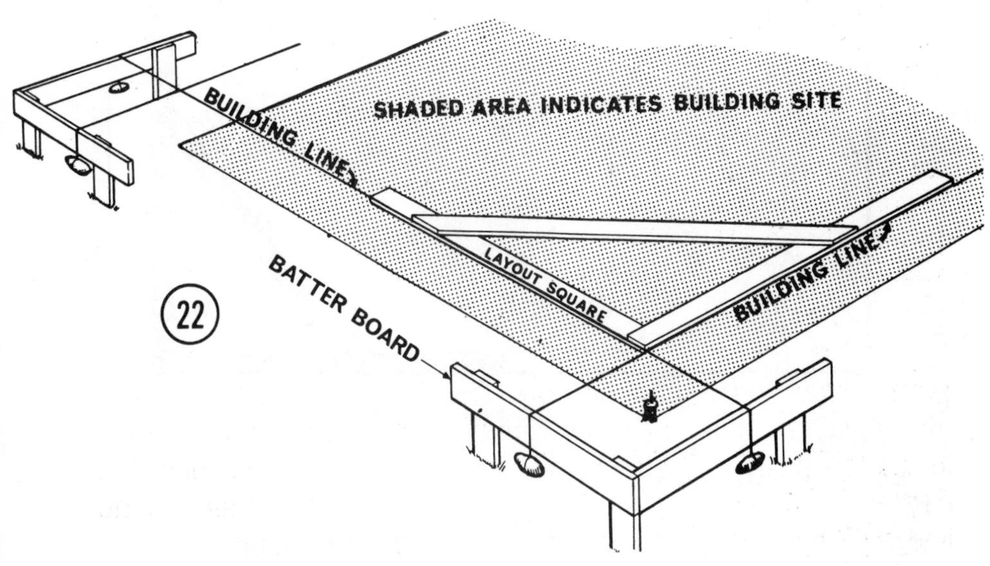

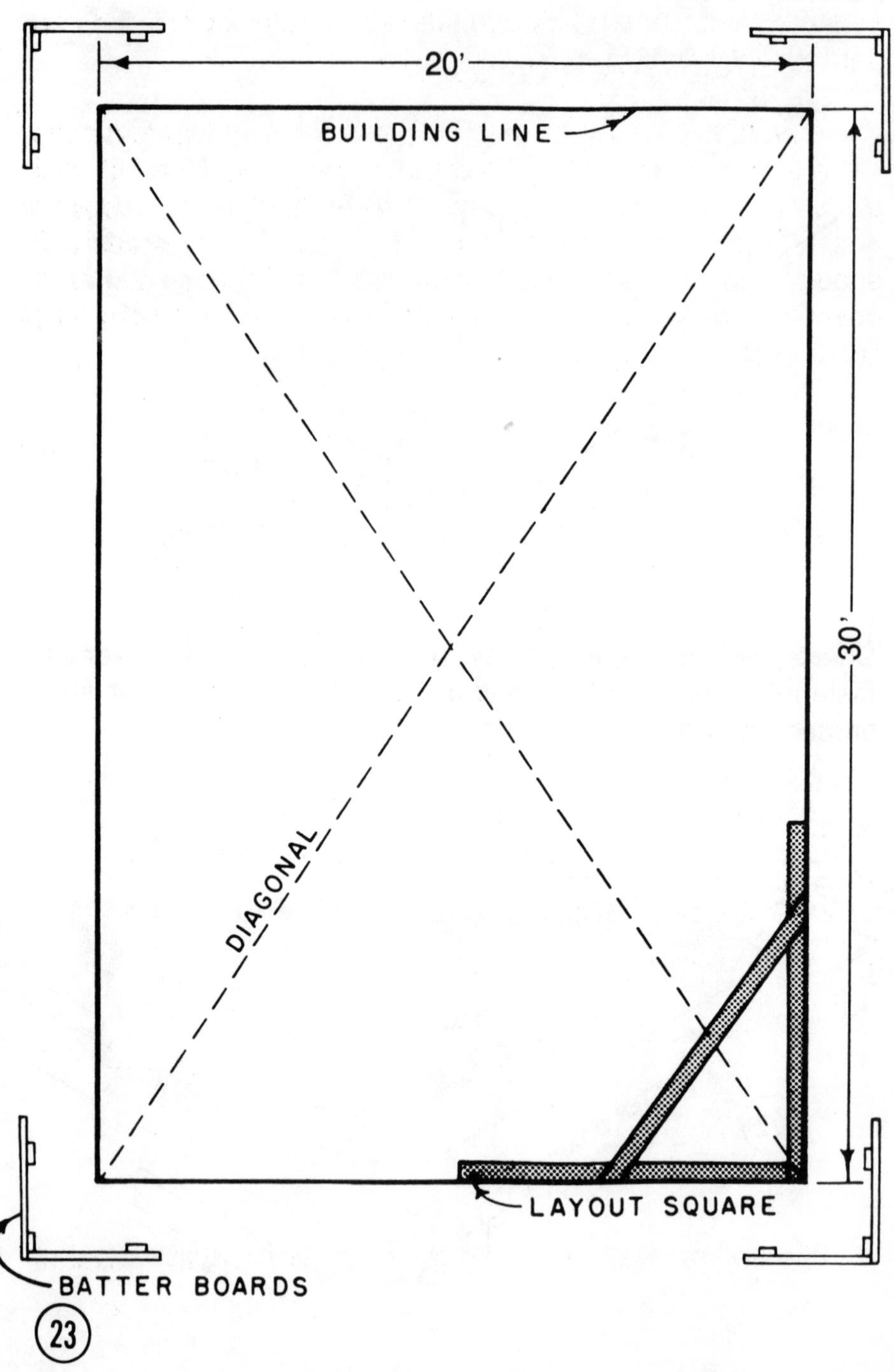
20'
BUILDING LINE
30'
DIAGONAL
LAYOUT SQUARE
BATTER BOARDS
23

If one person holds layout square against first line, another can sight and fasten line to batter board. When all lines have been leveled, check length of diagonals with a steel tape. If lines are square, diagonals will be equal length.

If you need to dig foundation trenches more than two feet deep, dig a 2 ft. wide trench. For less than two ft. depth, a 16" wide trench is ample. Run additional guide lines to dig trench width desired. Erect forms for footings 12" wide. This provides a 3" shoulder on both sides of 6" block, Illus. 24. Footings should go down below frost line and be 3½" thick, or thickness local building codes require.

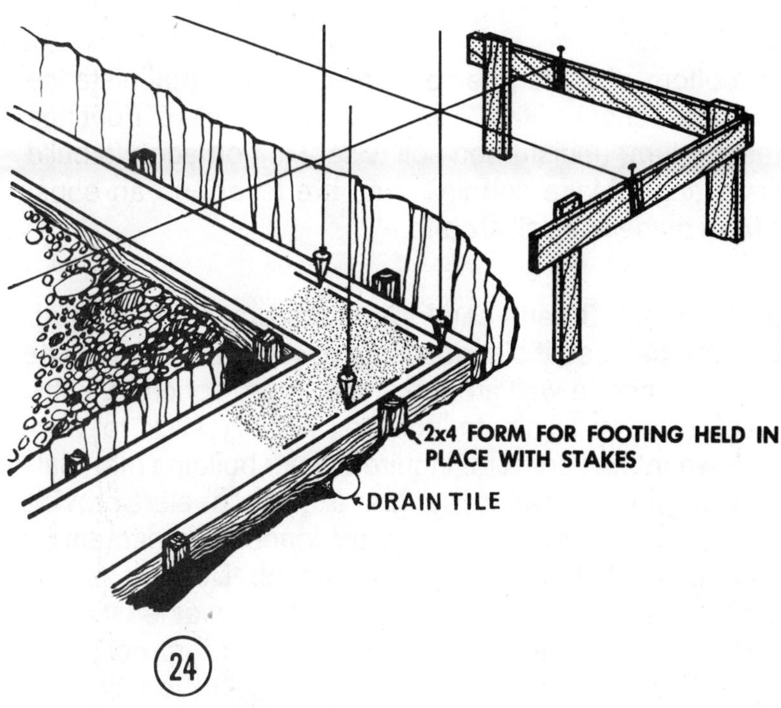

Pile and save all top soil for re-use. It's valuable—save all.

MATERIALS—FOOTINGS, FOUNDATION

Exact amounts of cement, sand, gravel or total number of concrete blocks depend on depth and thickness of footings, and height of foundation wall required by local building codes. As a starter, order:

10 bags of cement
5 yds. of sand
5 yds. of ¾" or run-of-bank gravel
½" x 12" machine bolts, nuts, washers
Approx. 74-6" x 8" x 16" concrete blocks per course below grade (due to breakage, order a few more blocks than actually required).

To insure bottom of trench being level, and an equal distance from guide lines, check with a 2 x 4 and level, Illus. 18. Footings should rest on firm, undisturbed soil. Use 2 x 4 on edge to build forms for footings. Make certain forms are level and an equal distance from guide lines all around.

Use 1 part cement, 3 parts sand to 5 parts gravel for footings. Fieldstone can be used if placed in bed of concrete. If fieldstone is used, flush concrete well around stone. Keep concrete a little on the wet side to eliminate pockets. Use a stick to work concrete down into all corners. Inquire at your building materials dealer for best prices on sand and gravel. Many dealers provide special discounts for five to seven yard loads. Buy cement as you go along. Don't store quantities on the job. Dampness may get in and harden it. Stack all materials on planks at least 8" off the ground. Cover securely with polyethylene, waterproof paper or tarpaulin. Stack lumber according to size and length. This speeds handling when the job starts.

Illus. 25 shows the 6" x 8" x 16" concrete blocks below grade. Illus. 10 shows top course with openings for doors. Foundation blocks can be 8", 12" or 16" or height above grade desired. Perimeter foundation projects above floor, Illus. 26.

25

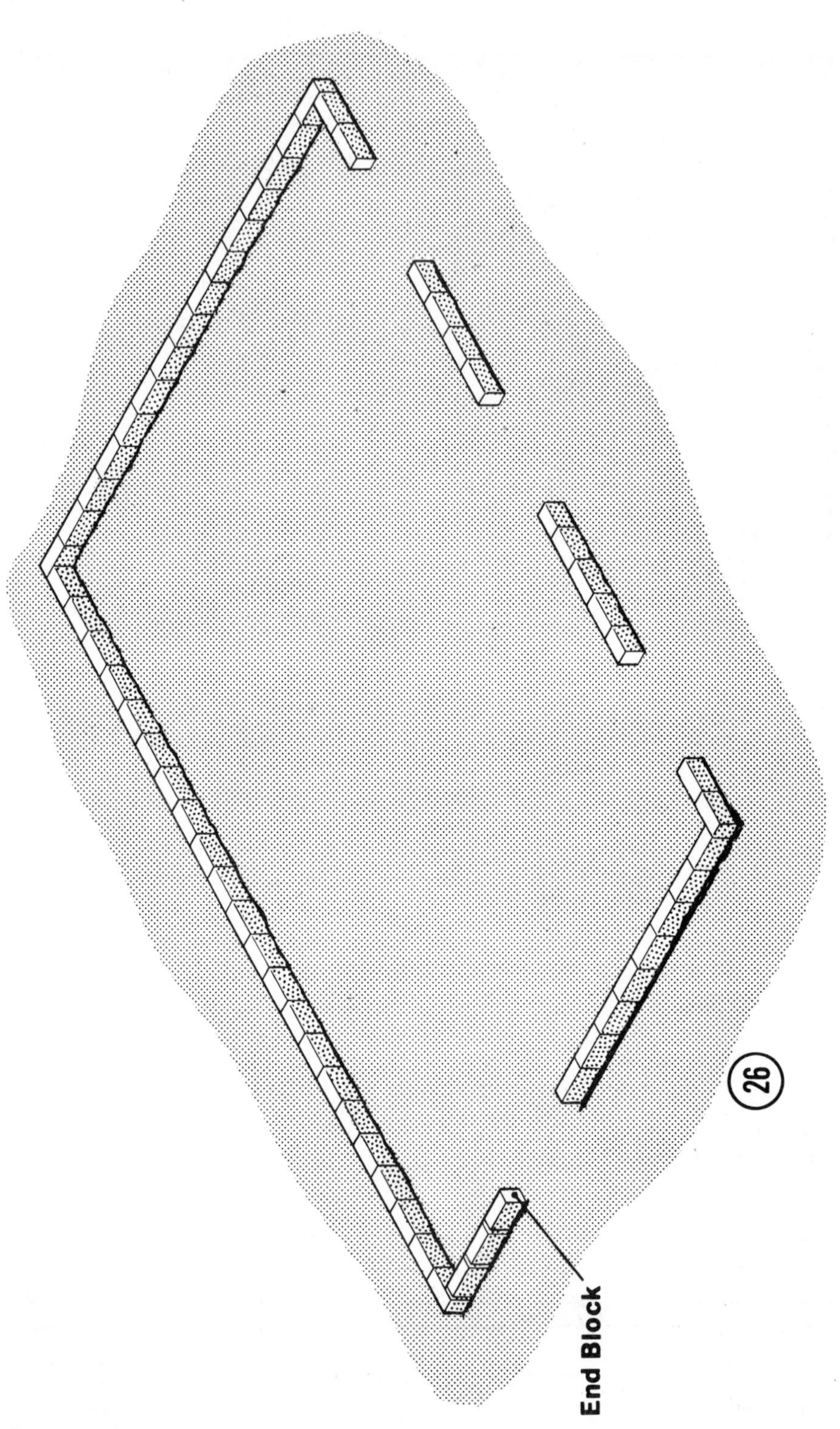
End Block
26

If you plan on running a water line or underground electric service, decide where these lines will enter the stable. To provide an opening, lay a 4" diameter drain tile or a piece of leader pipe, across footing, Illus. 27, below frost level. Mark location of opening on foundation plan.

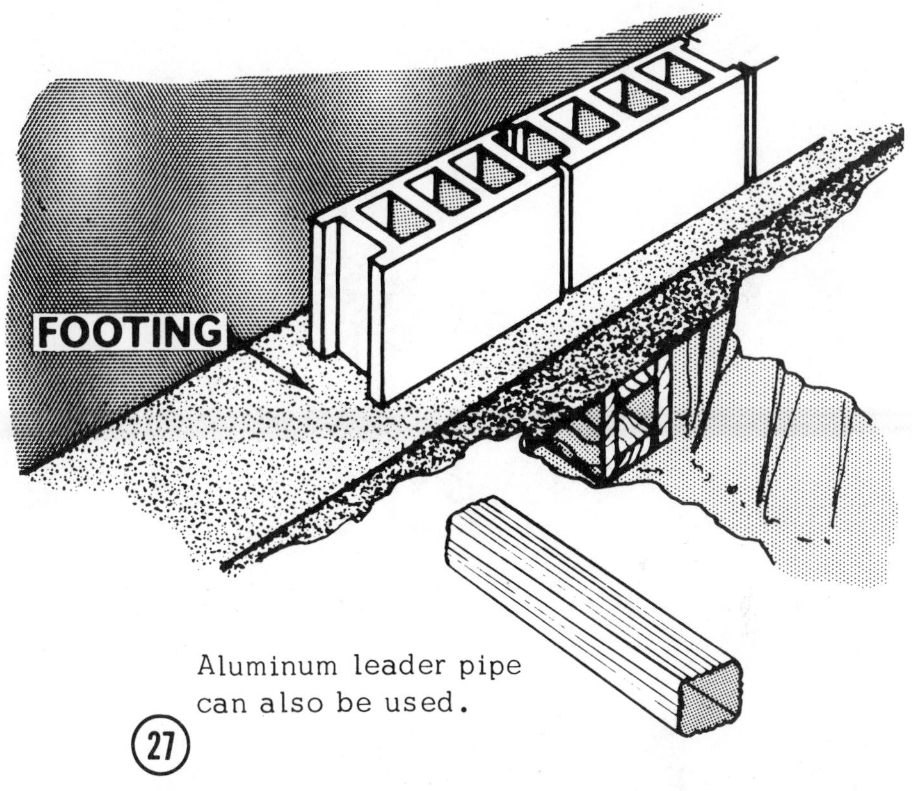

Aluminum leader pipe can also be used.

27

Allow footing to set at least two days before laying concrete block. Use 1 part cement, ½ part mason's lime to 6 parts sand for mortar to lay blocks.

Spread an inch thick bed of mortar on top of footing at both ends of one wall. Lay bed slightly wider than width of a block and length of two blocks. Drop a plumb bob from guide line, Illus. 24. Allow bob to mark mortar at corner, again two feet from corner. Set a corner block level and square at each end. Measure overall distance.

Now stretch a line along the top outside edge of these two blocks, Illus. 28. Hold line taut by wrapping it around a brick placed temporarily in position on top of corner block.

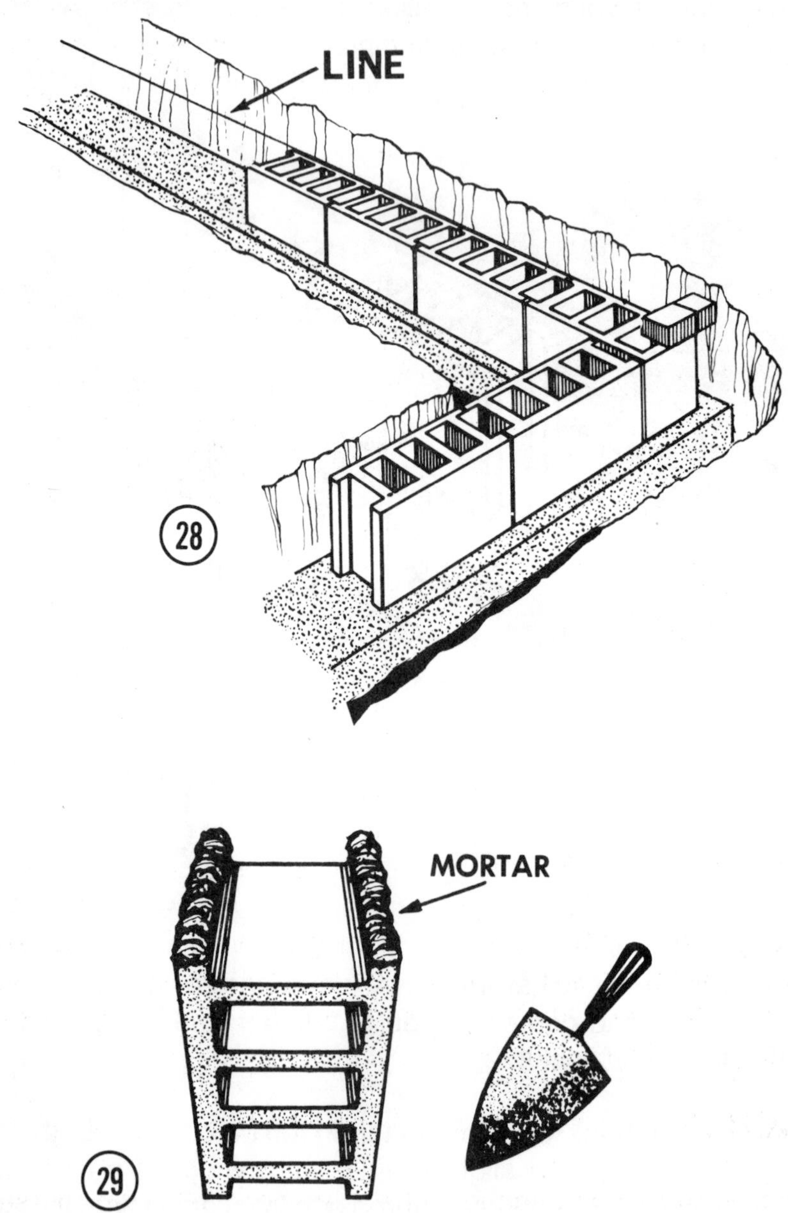

Butter up end of each block, Illus. 29.

Tap each block into place and check with a level and straight edge. Remove surplus mortar. Always lay a mortar bed as you go along. As you develop skill laying block, you can lay a larger mortar bed. Always stagger joints in each course, Illus. 30. Never allow vertical joints to line up on adjacent courses. Allow mortar in joints to set up a little, then use a jointer to finish joints.

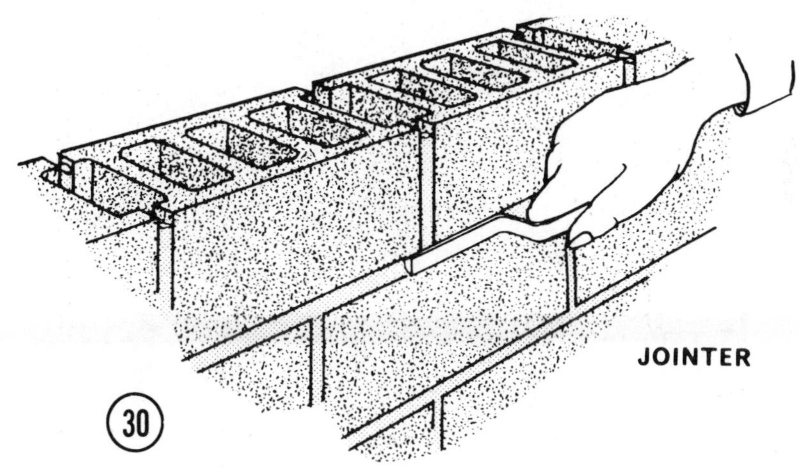

One course of 6 x 8 x 16" blocks above grade around perimeter of building, Illus. 8, will raise ceiling height 8", two courses 16". Setting block takes practice. You gain valuable experience working alongside an experienced mason. Use a jointer, Illus. 30, to finish vertical and horizontal joints. Lay corner blocks at ends and at door openings in position indicated, Illus. 9, 10.

ANCHOR BOLTS OR ANCHOR CLIPS

Position and embed anchor bolts or anchor clips in top course of perimeter foundation in position shown, Illus. 31. Use ½ x 10" or ½"x12" bolts with two washers. Bury one washer in concrete at head of bolt. Use other washer under nut, Illus. 32.

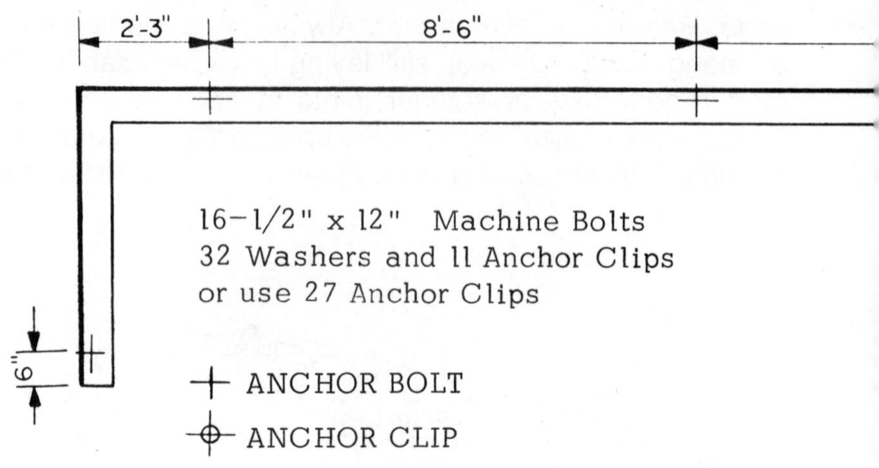
2'-3"
8'-6"
16–1/2" x 12" Machine Bolts
32 Washers and 11 Anchor Clips
or use 27 Anchor Clips
6"
ANCHOR BOLT
ANCHOR CLIP

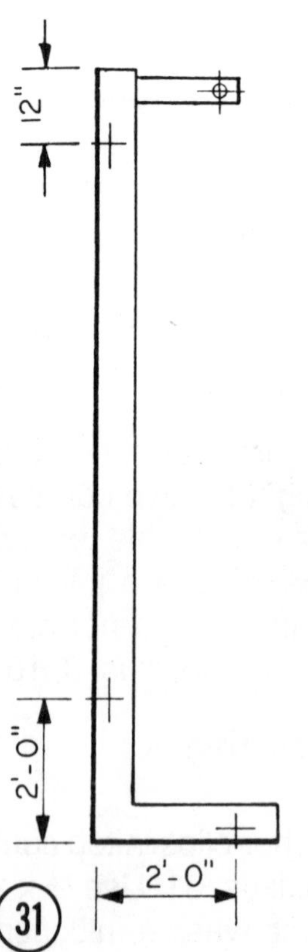
12"
2'-0"
2'-0"

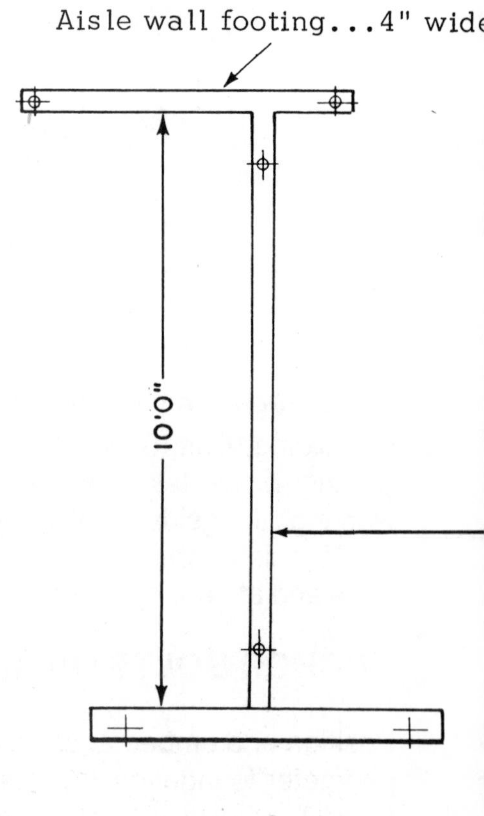
Aisle wall footing...4" wide
10'0"

31

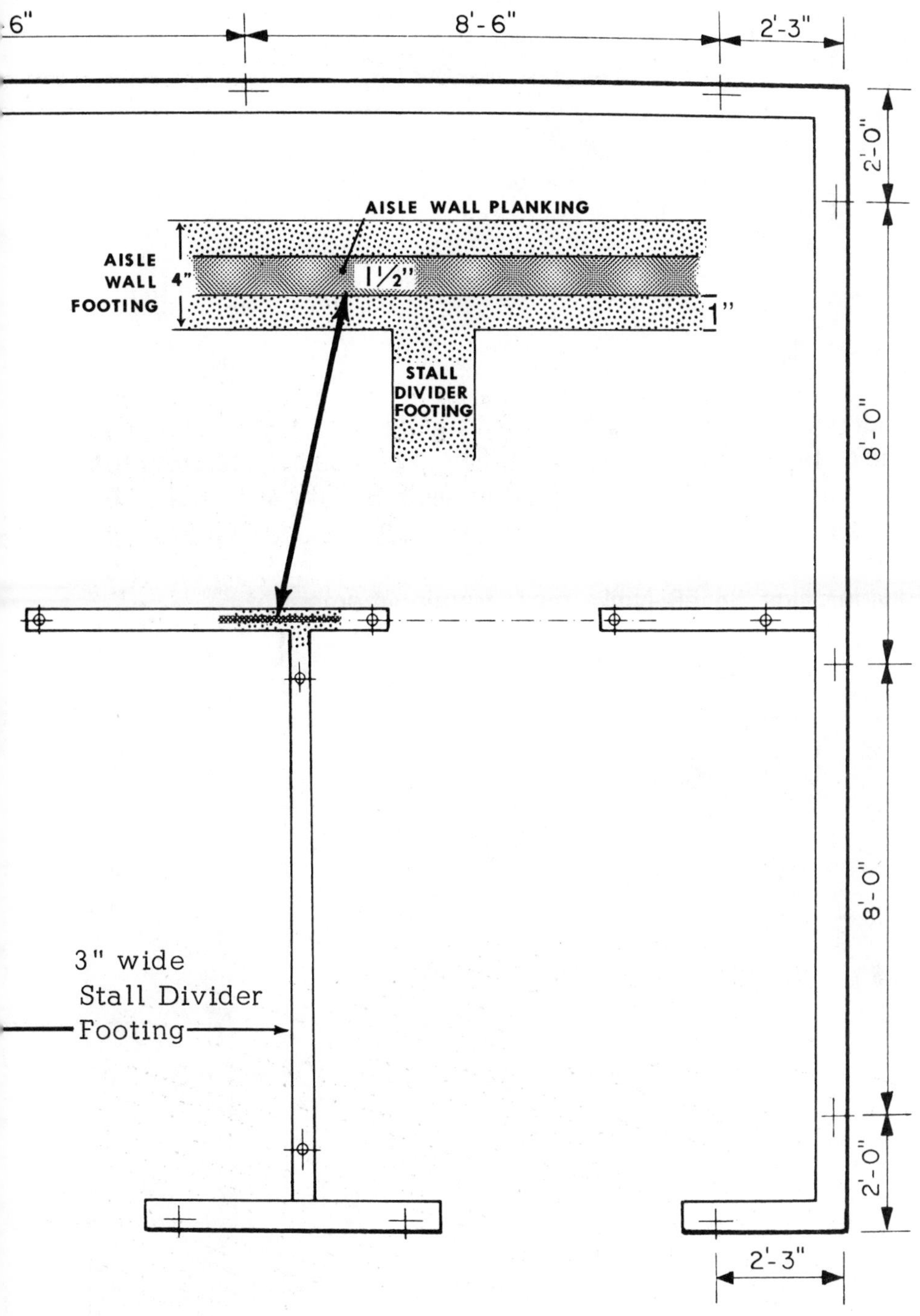

6"
8'-6"
2'-3"
2'-0"
AISLE WALL PLANKING
AISLE
WALL
FOOTING
4"
1½"
1"
STALL
DIVIDER
FOOTING
8'-0"
8'-0"
3" wide
Stall Divider
Footing
2'-0"
2'-3"

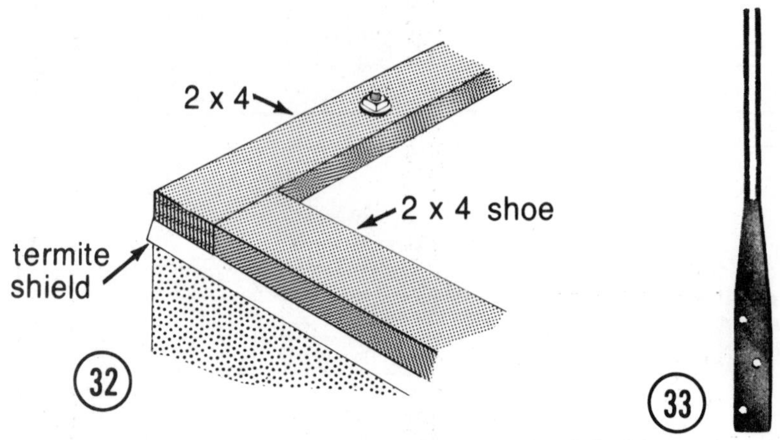

Anchor clips, Illus. 33, are easier to install. Just bend feet to right angle. Bury in concrete as you fill top course of block. Spread the arms and fasten clips to shoe, Illus. 34. These can also be used to anchor door frames to concrete blocks above grade, Illus. 35; and to posts B and bottom plank in each stall wall and divider, Illus. 36, 58. Nail arms to framing member.

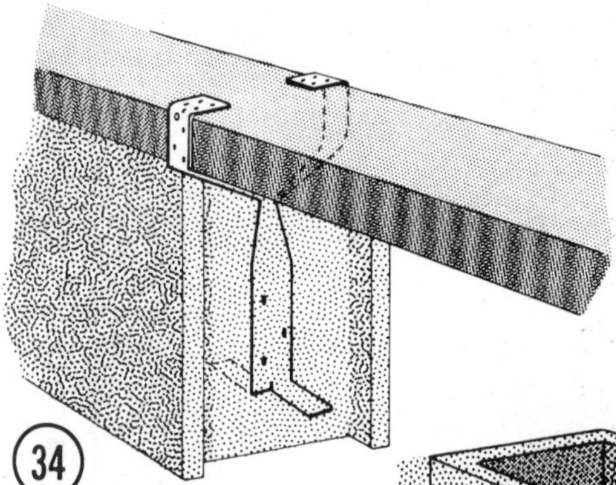

34

35

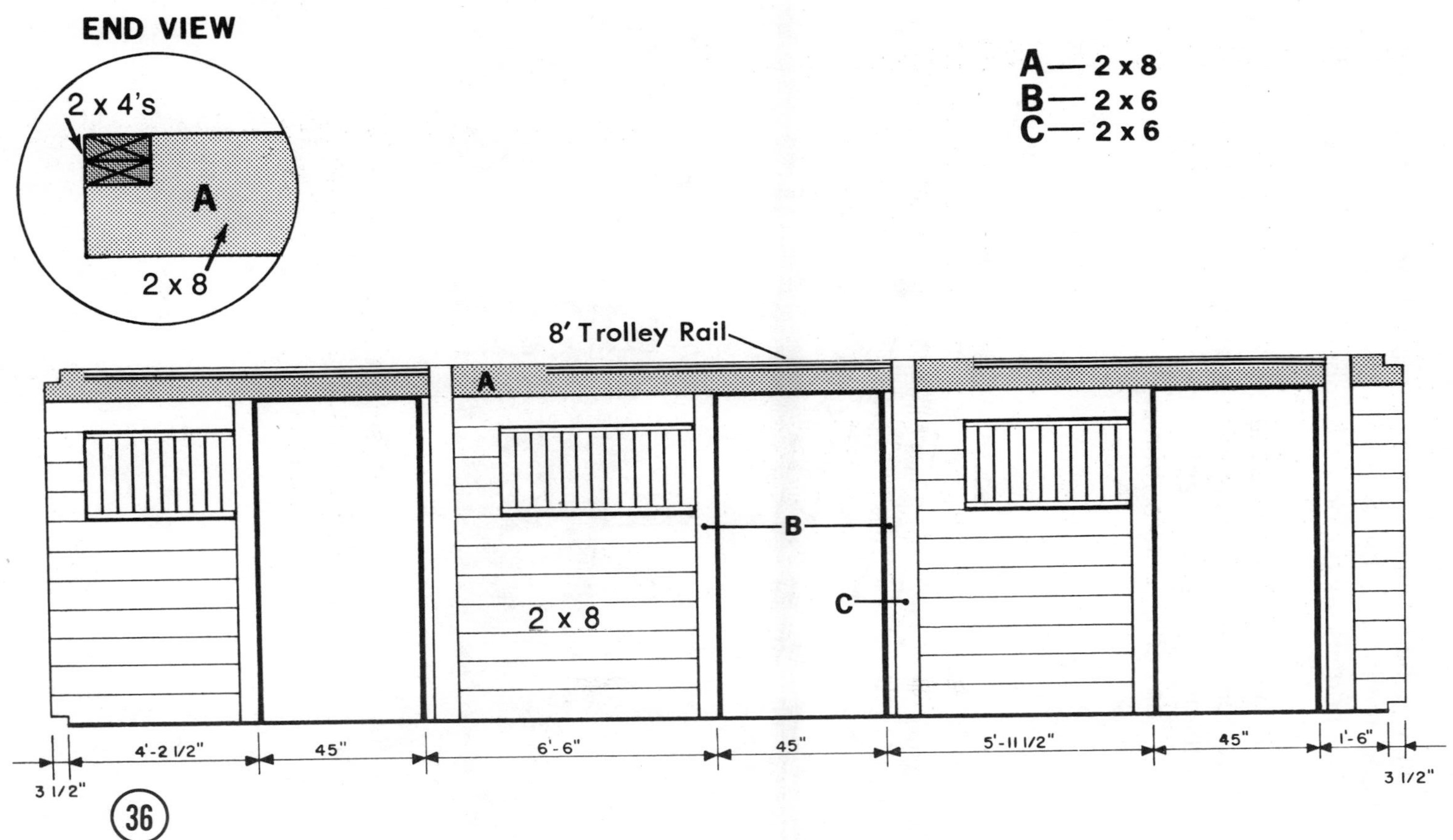

END VIEW
2 x 4's
A
2 x 8
A— 2 x 8
B— 2 x 6
C— 2 x 6
8' Trolley Rail
A
B
C
2 x 8
3 1/2"
4'-2 1/2"
45"
6'-6"
45"
5'-11 1/2"
45"
1'-6"
3 1/2"
36

When you have raised building more than 8" on foundation, embed anchor clips between end blocks in door openings, Illus. 35. This permits anchoring 2 x 6 door frame to foundation.

Measure 3¼" from inside edge of block and snap a chalk line. Embed bolts or clips in concrete along this line. If you use bolts, allow them to project 3" above block, Illus. 37. If you place a washer on bolt, before burying, it provides even greater anchorage.

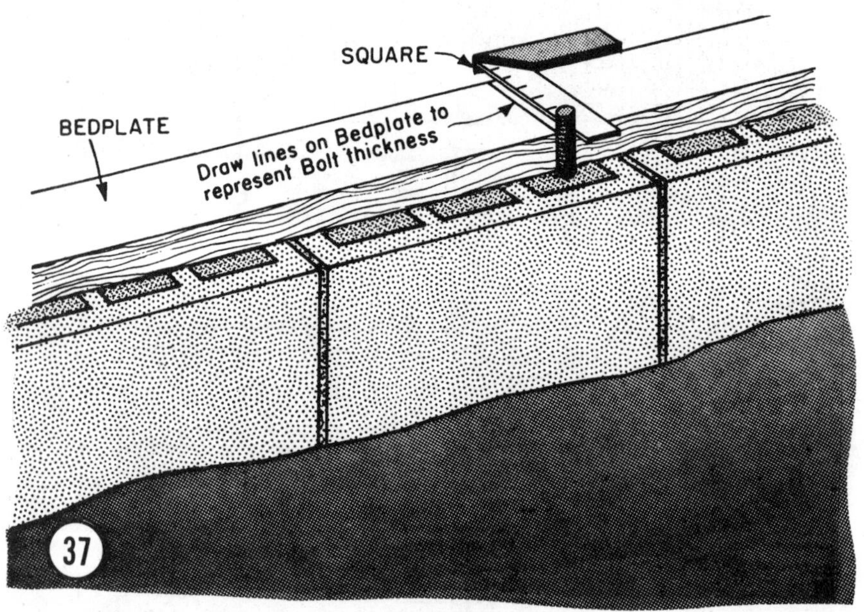

Allow foundation blocks to set at least three days before backfilling around outside of foundation.

FOOTINGS—AISLE AND STALL DIVIDER WALLS

Since the footing for each stall divider and aisle wall, Illus. 31, is in a protected area, footings need only go down 8", 10" or 12" to undisturbed soil. Use 2 x 8 stall planks for footing forms, Illus. 38. Use stakes to hold forms in place, Illus. 39.

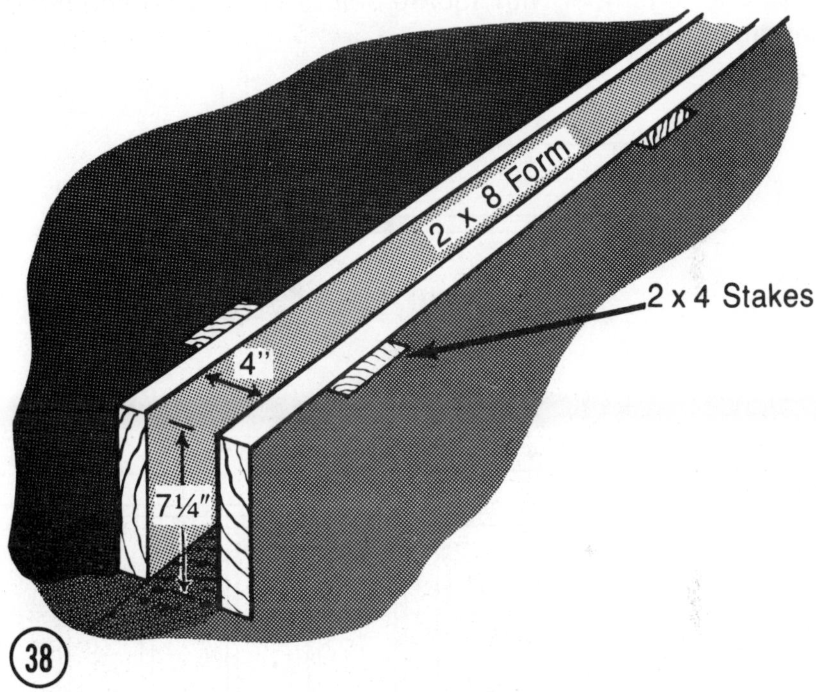

38

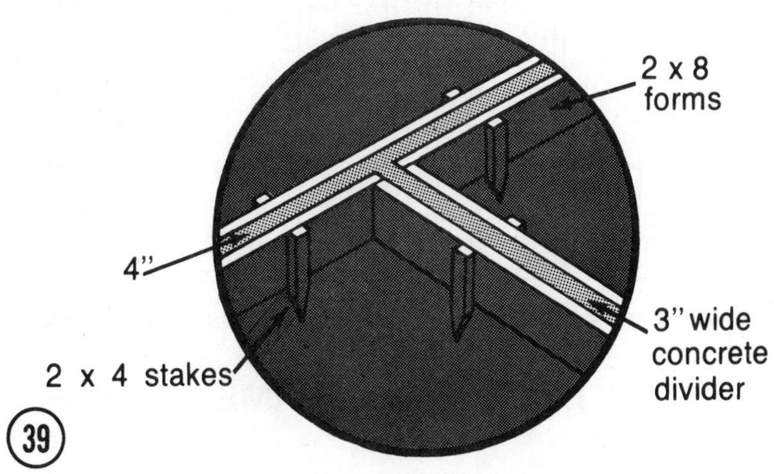

39

WALL FRAMES

Due to variance in lumber width and thickness, carefully measure length of each inner framing member against construction before cutting. Maintain overall dimensions shown. Always select the straightest 2 x 4's for studs, plates and shoes. Check end of lumber with square before measuring and cutting to length required, Illus. 40.

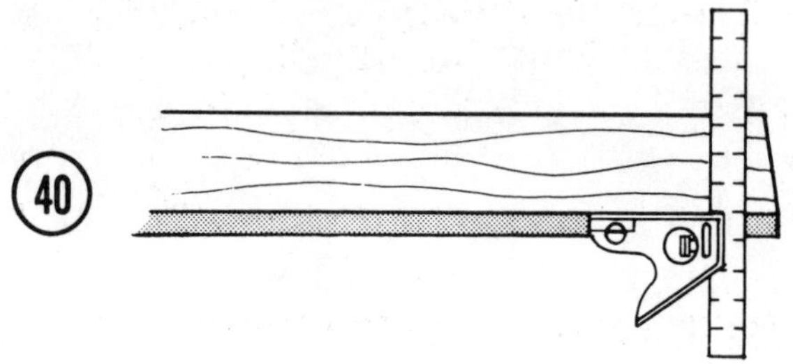

3'-0" 3'-4" 7'-4" 2'-9 1/2

Space Stud Where Required

20" 16" 16" 16" 8" 8" 16" 16" 16" 16" 16" 16"

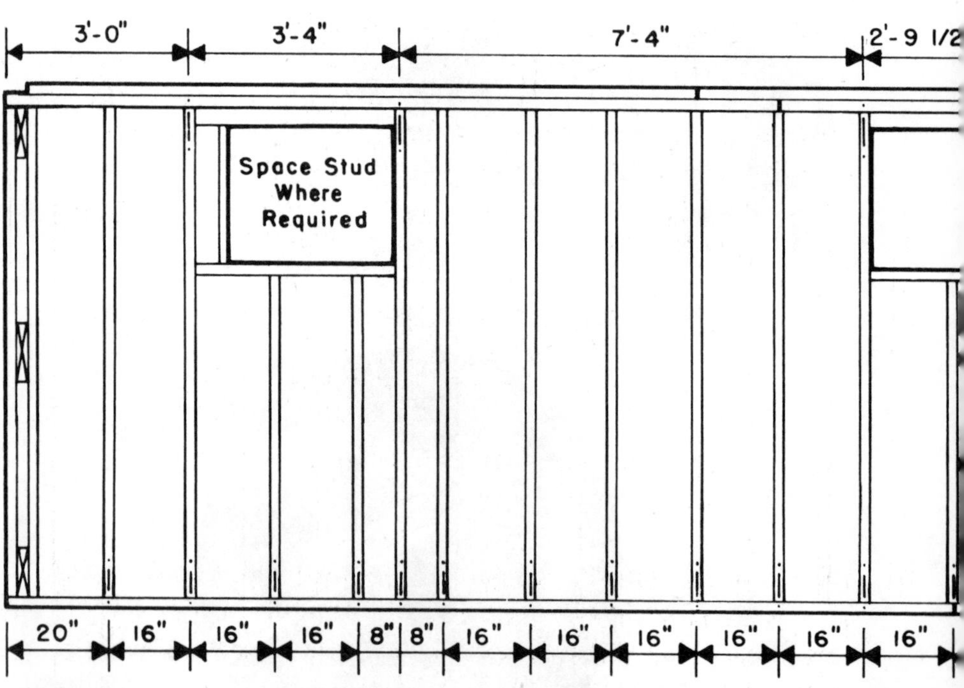

41 LEFT WALL FRAME

If you don't own an electric handsaw, table or radial arm saw, buy or rent one. This one job will make the purchase worthwhile. It's like having an extra man on the job who always makes square cuts.

When two or more lengths of 2 x 4 are required for a shoe or plate, butt ends over, or under a stud, never over a header, Illus. 41. Never butt ends of a plate over same stud used to cover ends of a shoe. Stagger the joints.

Shoe — 2 x 4 — 2/16'
Plates — 2/8', 2/12', 2/14'
Studs, Headers, Sills — 30/8'

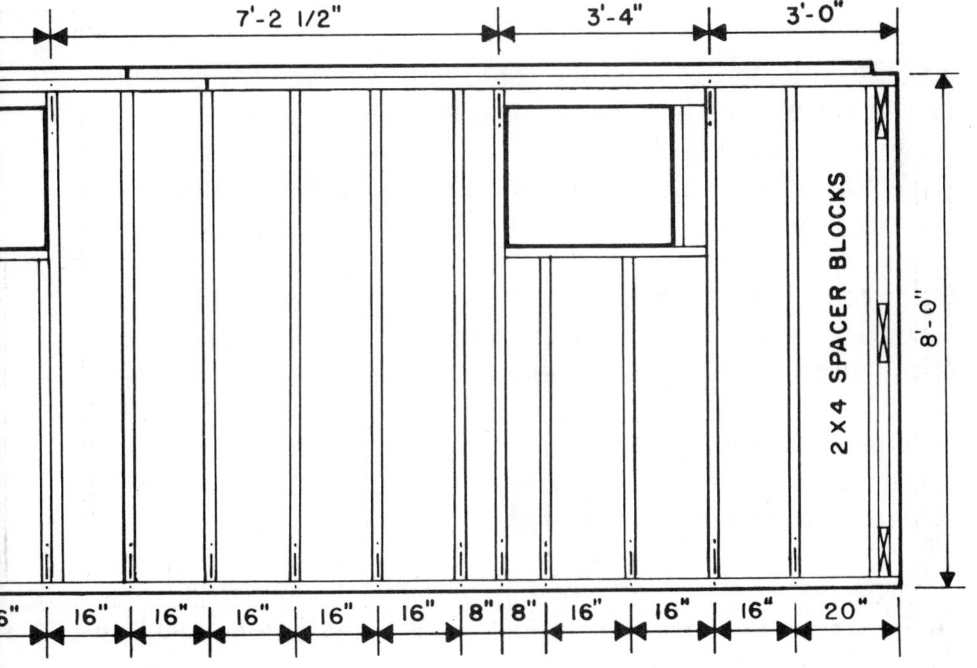

Select a level area to lay out a wall frame. Always place a plate and shoe together on a flat surface, Illus. 42, then measure 16", or distance required, and mark exact location for each stud. Using the square, draw a line across the edge of both the plate and shoe to indicate center of stud. This simplifies assembly.

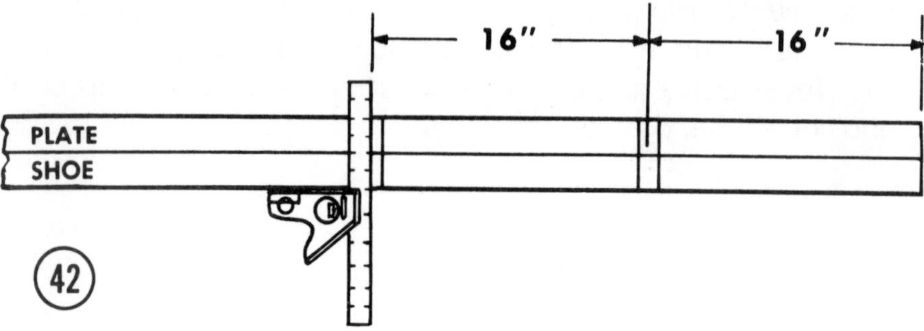

Since window manufacturers specify rough opening size for the windows they manufacture, always frame openings to size recommended. All openings for stall windows should be roughed in, in position shown.

Nail the plate and shoe to each stud with two 16 penny common nails at each joint. Use 8 penny nails when toenailing.

If building code requires using a double 2 x 4 shoe, and/or plate, cut length needed to maintain overall dimensions indicated.

Side, front and back wall frames are assembled on a flat surface, then raised and braced in position. Before standing on, or working near any framing member, be sure it is secure.

Framing illustrations indicate openings for stock size windows. If you make any changes, your dealer can suggest rough opening size required for window selected.

After assembling a frame, measure length of diagonals, Illus. 43. Use a 50 ft. steel tape. If diagonals aren't equal in length, frame isn't square. Strike frame at corner to square it up. When square, nail a 1 x 6 temporary brace diagonally across frame, or nail wind braces. Note directions on page 56.

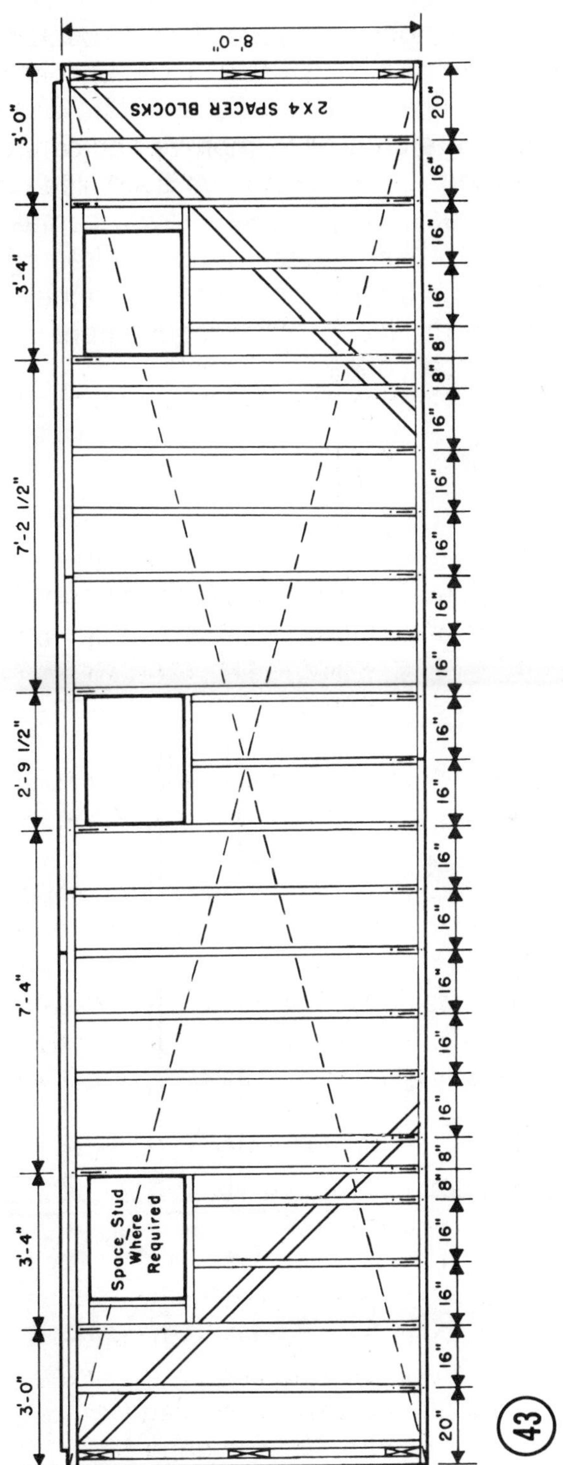
8'-0"
2X4 SPACER BLOCKS
3'-0"
3'-4"
7'-2 1/2"
2'-9 1/2"
7'-4"
3'-4"
3'-0"
20"
16"
8"
Space Stud Where Required
43

Illus. 44 shows the right wall frame; Illus. 45 back frame; Illus. 46 front frame.

Building codes were created to improve and control methods of building. Most codes were drawn up by skilled craftsmen who knew their business. Some were drawn to make work for a chosen few and to bluff others from being free agents. Always find out whether the framing proposed for the building you want to construct meets local code requirements. Always alter framing to meet local codes.

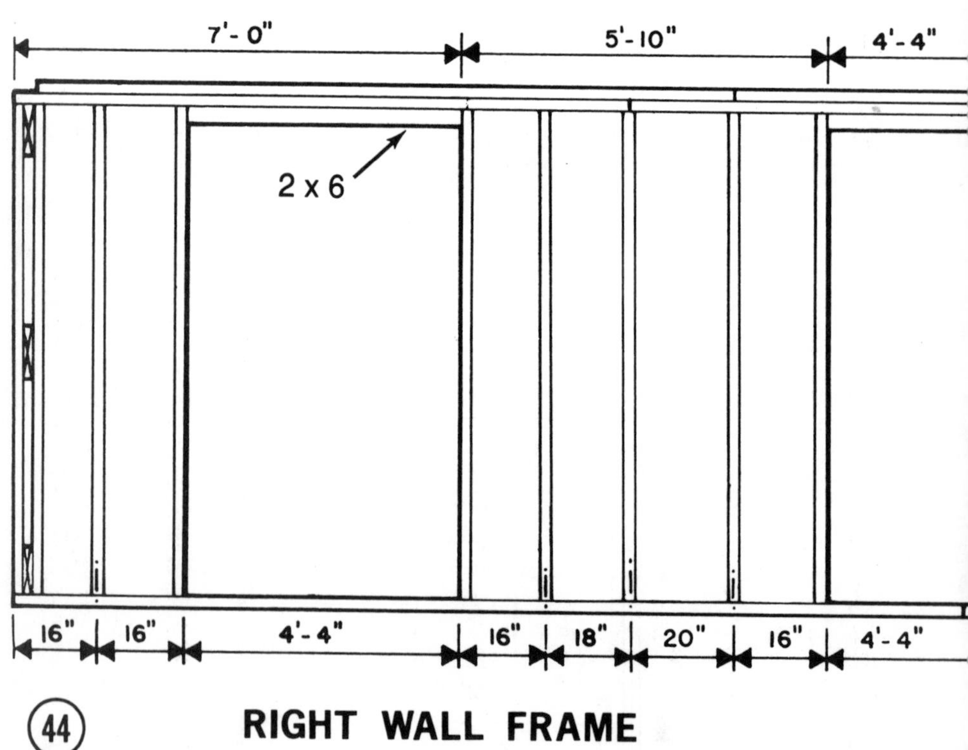

(44) **RIGHT WALL FRAME**

If the code specifies a double 2 x 4 shoe, or sill, as it's called in many areas, put it in. If a building inspector gives you a lot of rhubarb about an amateur not being able to build like a "pro", take along a tape recorder to get a permanent record of his comments. While most people in public office are honest, those seeking a handout frequently find a lot of fault until they feel the green. Start recording a building inspector's comments, "so you won't forget anything he says" says you, and you'll be amazed how little he has to say, and how quickly he stops reading the riot act.

Shoe — 2 x 4 — 2/16'
Plates — 2/8', 2/12', 2/14'
Studs — 18/8'
Headers — 3 — 2 x 6 x 10'

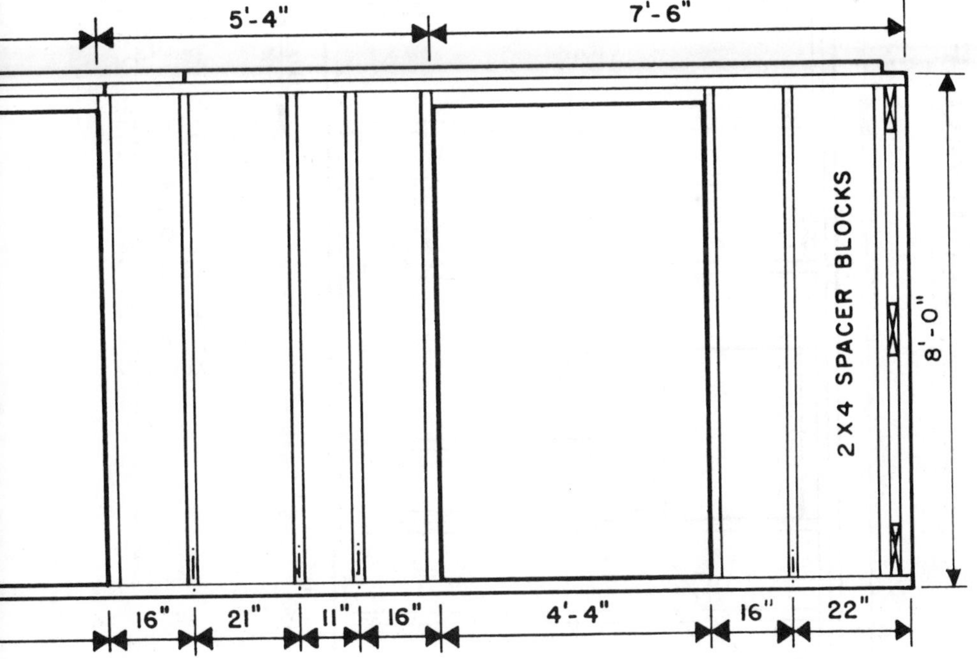

BACK WALL FRAME

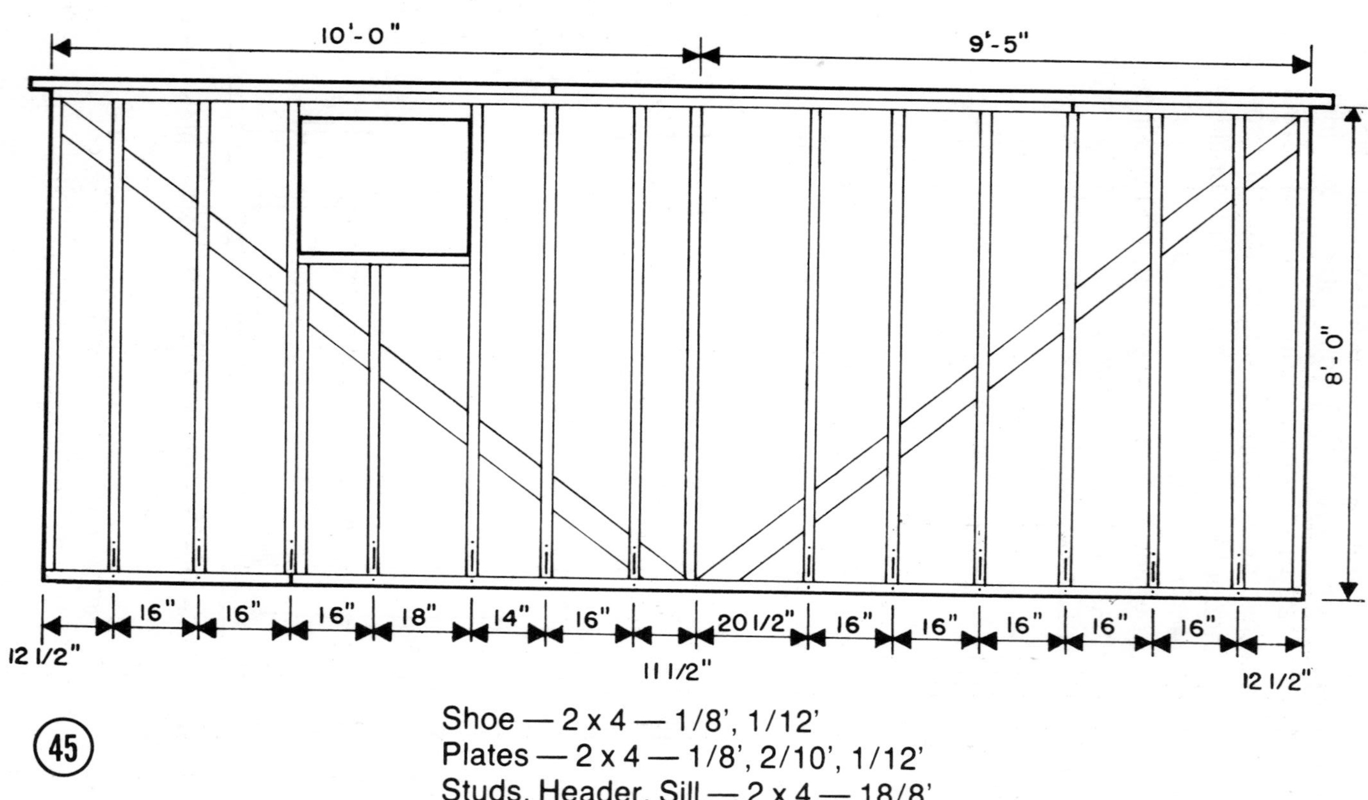

Shoe — 2 x 4 — 1/8', 1/12'
Plates — 2 x 4 — 1/8', 2/10', 1/12'
Studs, Header, Sill — 2 x 4 — 18/8'

Plus 1 x 4 wind braces.

FRONT WALL FRAME

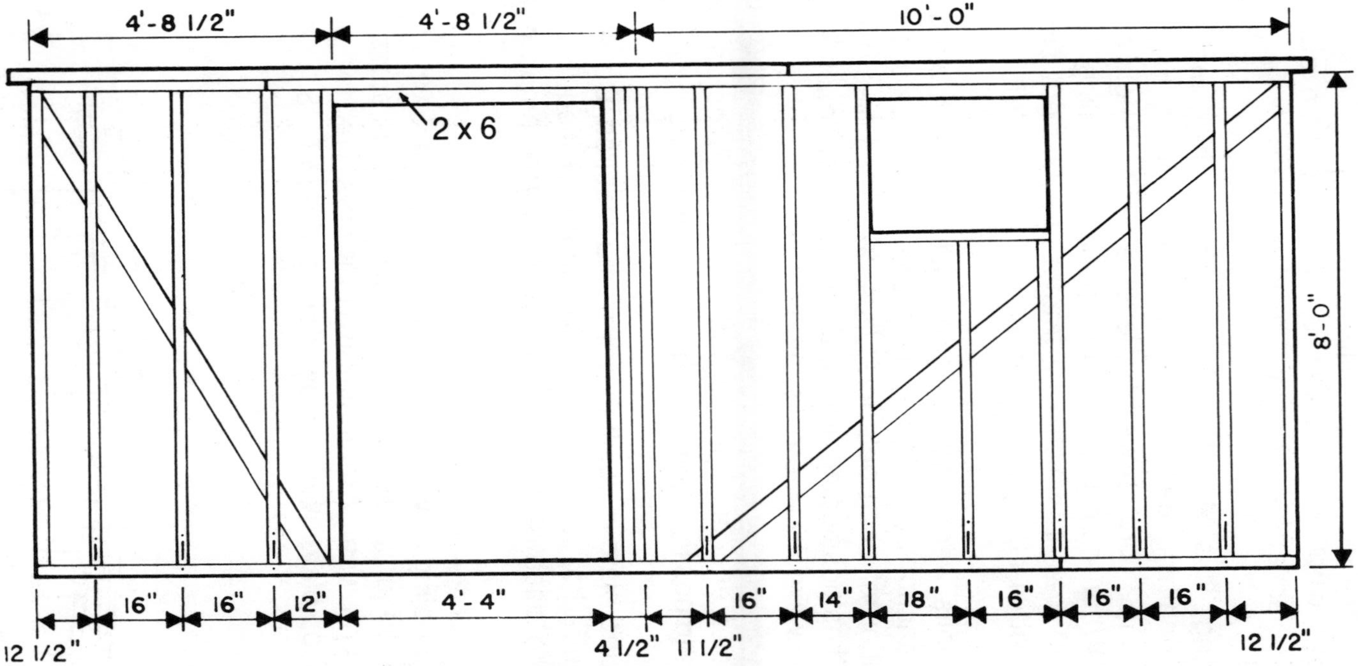

Shoe — 2 x 4 — 1/8', 1/12'
Plates — 2 x 4 — 1/8', 2/10', 1/12'
Studs, Sill — 2 x 4 — 18/8'
Headers — 2 x 6 — 1/10'

Plus 1 x 4 wind braces.

If any change in framing stumps you, your building material retailer can explain it. Most building material retailers are real honest Joes who know what goes. Consult them about any building problem and you, too, will build like a pro.

Wall framing, Illus. 41, 44, 45, 46 permits using most 4 x 8 plyscord sheathing panels without cutting. When frames are placed on an 8" curb above floor level, a ceiling height of 8'8" is achieved. When curb is 16" above floor, 9'4" ceiling height is gained. The way in which wall frames are laid out simplifies sheathing with 4 x 8 plyscord panels.

If anchor bolts were embedded in blocks, position each shoe alongside anchor bolts. Using a square, draw lines to indicate exact position of each bolt hole, Illus. 37. Drill holes oversize. This simplifies positioning frame. Be sure to mark each shoe so you will know exactly where it's to be placed.

If you embed anchor clips, spread the clips. Place and brace each frame in position then nail the clips to the shoe.

Lay shoe and plate in position wall frame requires. Cut studs to length required to maintain overall height specified. Nail shoe and plate to each stud using two 16 penny common nails at each joint.

When assembling a wall frame, nail headers and short studs where needed for window and door openings. Add studs where required if you want to nail stall wall dividers. Nail top plate in position so top plate overlaps lower plate, Illus. 47, 48.

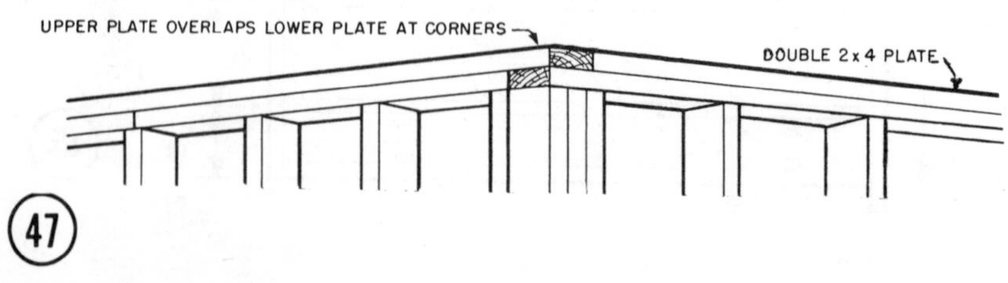

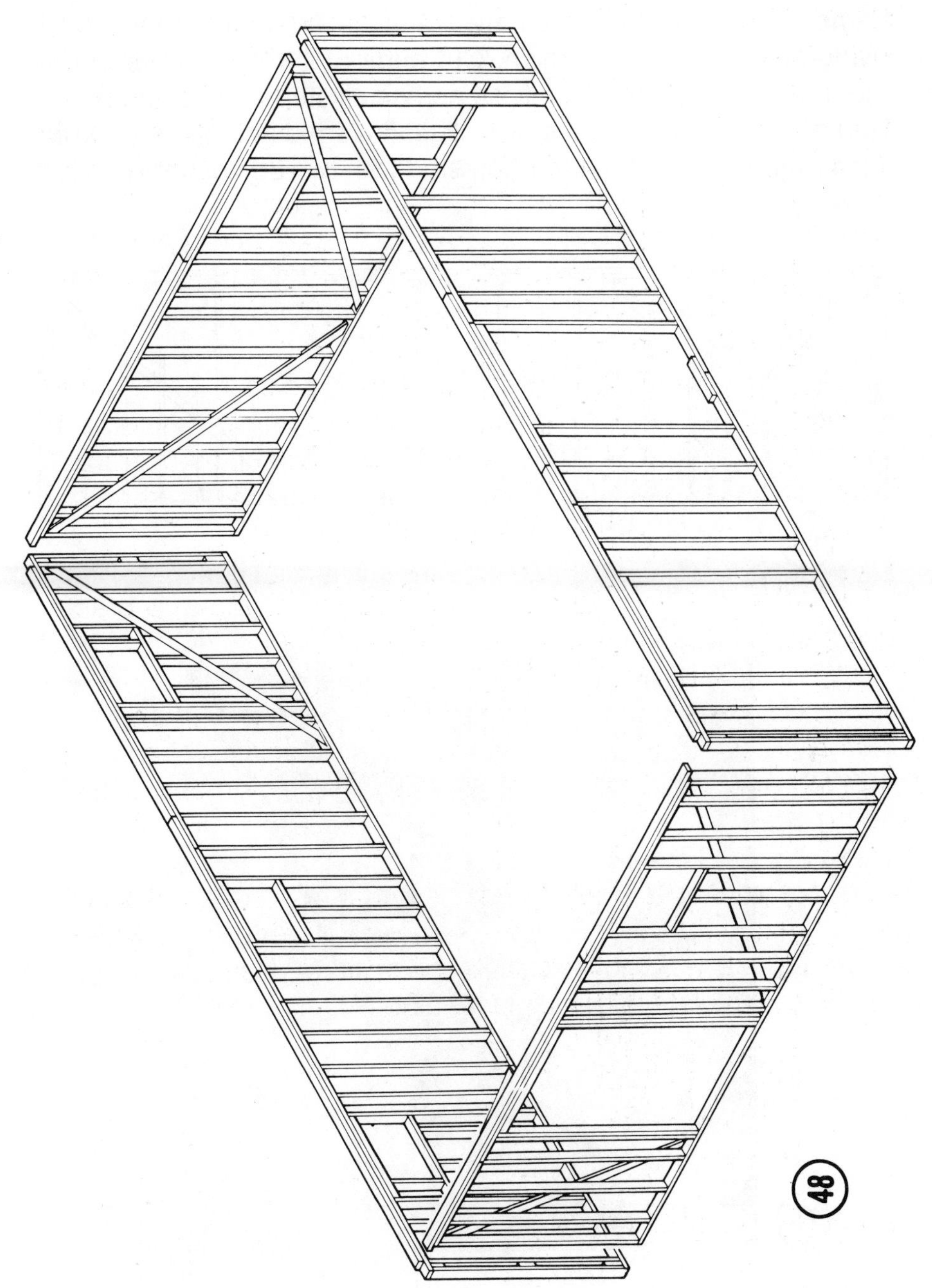
48

NOTE: In areas that experience high winds, many codes specify wind braces, Illus. 49. Use 1 x 4. After assembling a frame, and diagonals indicate frame is square, place 1 x 4 diagonally in position shown. Cut ends of 1 x 4 to angle frame requires. Draw position of 1 x 4 across all studs, shoe and plate. Remove 1 x 4 and notch studs, Illus. 50, shoe and plate to depth 1 x 4 requires. Nail 1 x 4 to each stud with two 8d common nails.

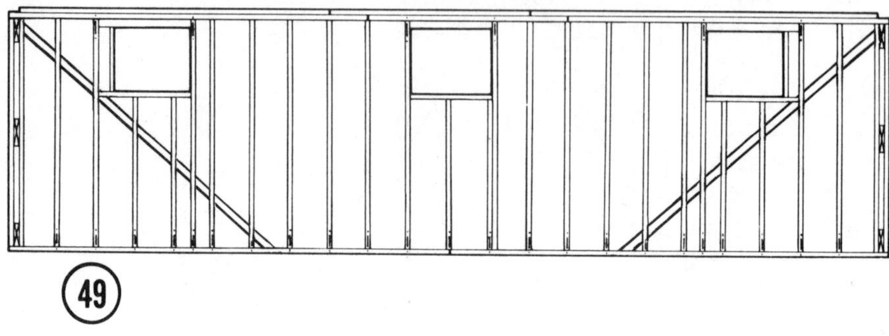

49

50

Follow same procedure in laying out and raising each wall frame. Get adequate assistance before moving or raising a frame. Don't strain yourself, or allow anyone else to do it. When frame is raised, nail temporary braces to stakes, Illus. 51.

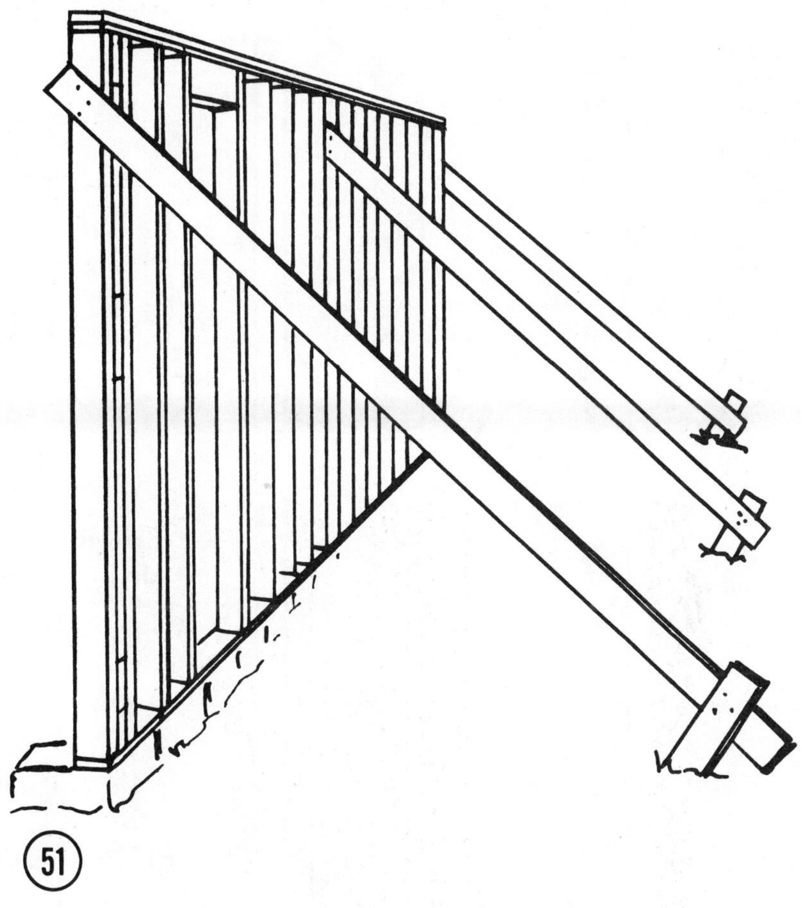

Raise, brace each frame. Drive 2 x 4 stakes into ground. Using a level, plumb each frame in two directions. Only hand tighten nuts on anchor bolts.

To ascertain whether frame is plumb, do this. Drive a nail into plate, Illus. 52, and tie a plumb bob line say 1" from end of plate. Now measure distance point of bob is from shoe. When point of bob measures 1" from shoe, end is considered plumb. Do the same over side of plate. Brace frame in plumb position.

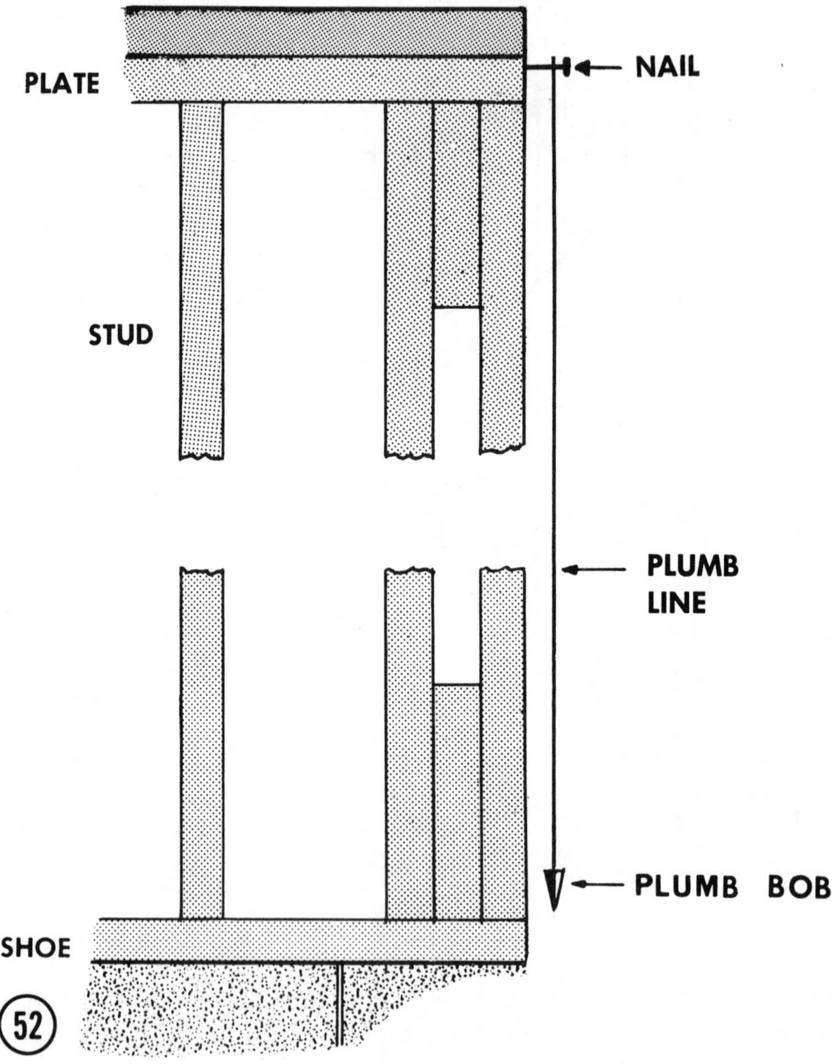

To make certain all frames are straight, drive a nail into ends of a plate and stretch a line, Illus. 53, from one end to other. Raise line at ends with 1 x 2. Use another piece of 1 x 2 to test space between plate and line. Straighten frame.

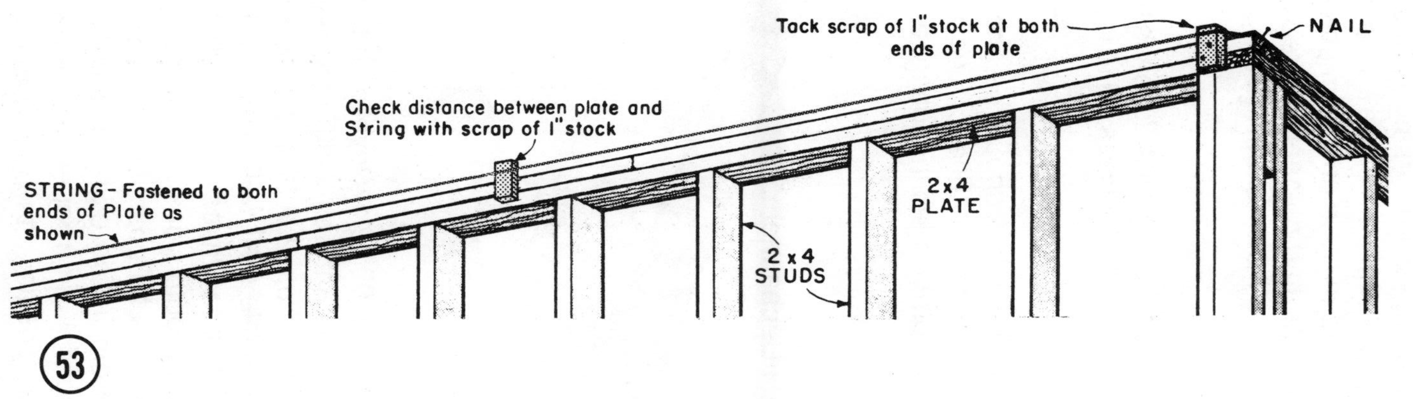
Tack scrap of 1"stock at both ends of plate
NAIL
Check distance between plate and String with scrap of 1"stock
STRING - Fastened to both ends of Plate as shown
2 x 4 PLATE
2 x 4 STUDS
53

When all four frames are raised, setting in position on bolts, only hand tighten nuts. Level and plumb one frame, then do the same to the others. If you had to raise and shim shoes with a piece of slate or shingle, mix one part cement to three parts of sand and flush this under shoe. Allow to set three days then tighten anchor bolts.

While the next framing step of construction is to erect the center aisle beam A, Illus. 36, it's first necessary to lay footings for aisle wall and stall dividers.

The aisle wall and stall dividers should be anchored to concrete footings. This can be done in one of two ways. Either with anchor clips, Illus. 33, embedded in footing; or with U-channels, Illus. 54.

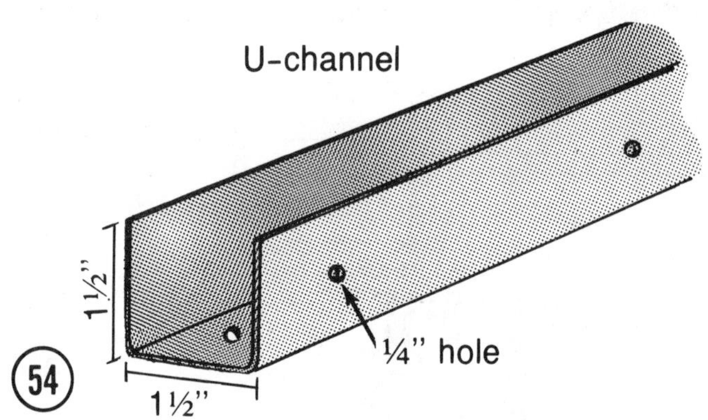

Dig footing trench 8" or 10" below level of dirt stall floor. To pour footings level, and to height and width required, do this. Dig an 8" wide trench and set up 2 x 8 or 2 x 10 forms 3" wide for dividers, 4" wide for aisle stall wall, Illus. 38, flush with level of dirt floor. Toenail forms to stakes. The 3" width allows a narrow protective shoulder on both sides of divider planks; the 4" width provides sufficient width for posts C and B. Stretch footing form guide lines 10'0" from inside edge of foundation blocks. Aisle wall should be placed 1" in from edge of footing. Bend feet on anchor clips and position them in approximate location indicated, Illus. 31.

Pour footings using one part cement, three parts sand, five parts ¾" gravel. Use a stick to eliminate air pockets. Position clips so they project above forms at least 6". Allow footings to set three days and remove forms and compact earth against footings. Anchor clips are nailed to both side of posts B, also to bottom board of stalls.

The other method of anchoring stall walls to footings is with U-channel, Illus. 54. After pouring footing level with forms, 1½" x 1½ x 1½" U-channel, cut to length specified, and predrilled with ¼" holes in bottom and sides, is placed in position on the freshly poured footing. Each length of U-channel is checked with a level and guide line.

When positioned accurately, a nail is pushed through each hole in bottom to mark concrete. The U-channel is carefully lifted out and an expansion shield, Illus. 55, with lag screw, is embedded in concrete where holes require. Allow concrete to set three days, then remove lag screw and anchor U-channel to expansion shield. It will be necessary to notch starting plank to receive head of lag screw.

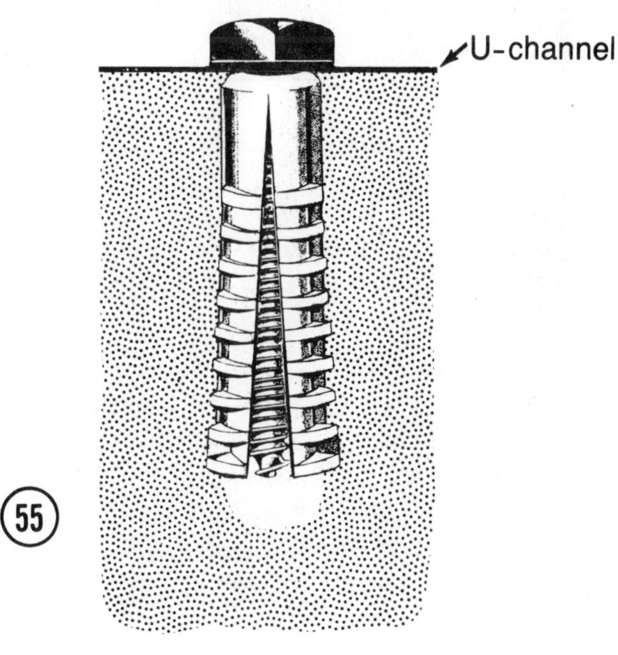

To assemble a center aisle partition beam, A, Illus. 36, so it butts together over a post B, select two straight 2 x 6 x 16. Saw ends square. Cut one 14'9", the other 15'3". Butt ends together. Toenail, then apply a 2¾ x 10" splice clip, Illus. 56, on stall side.

Notch ends, Illus. 36, 57, so top edge of beam is level with top of double plate. Place beam in position and nail it to stud with 16 penny nails.

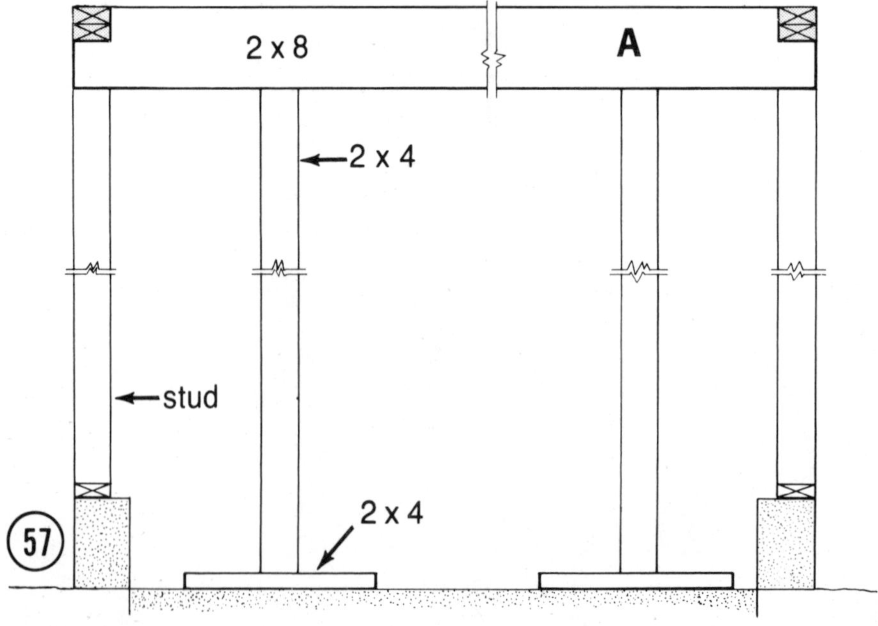

Drive a nail into end plates. Stretch a taut line along one edge of beam, Illus. 53. Raise line with 1 x 2 blocks at both ends. Test spacing between line and beam with another piece of 1 x 2. This will show you where the beam is low or high. Support beam in level position with temporary 2 x 4 posts.

When beam is level, cut 2 x 6 posts B, Illus. 58, to exact length required. Toenail posts B to beam. When B reads plumb, fasten U-channel or anchor clip to bottom of B. Nail a 3⅜" x 5" truss clip to A and B on aisle side of AB, Illus. 59.

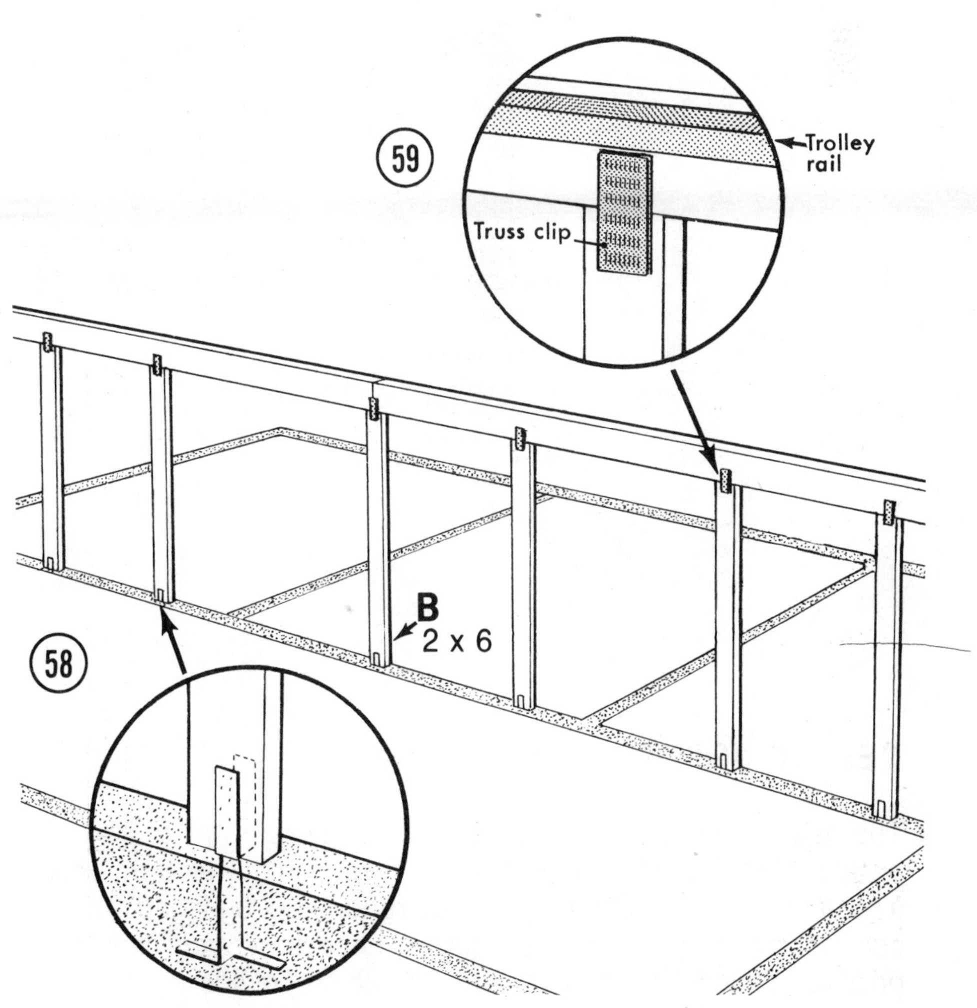

Space posts B to allow approximately 45" for door opening, Illus. 36.

To further secure posts B in position, and to provide a sliding door stop, cut 2 x 6 posts C to full height of beam, Illus. 36. Spike C to B and A in position so C provides a 1½" recess for door. Balance of C covers stall boards. If you use a U-channel, it will be necessary to notch bottom end of C—⅛ x 1½ x 5½ so it receives side of U-channel, Illus. 60.

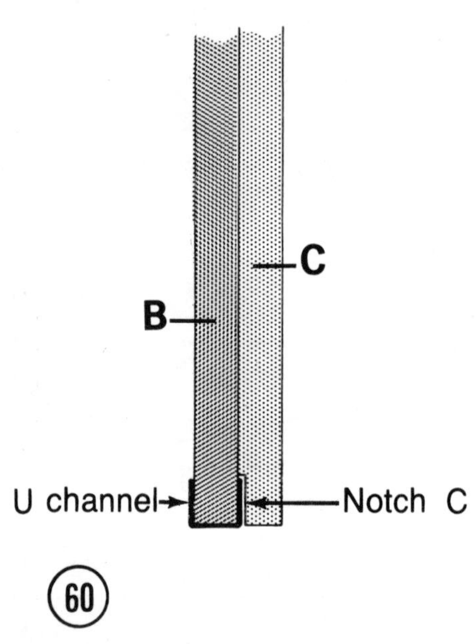

60

CEILING JOISTS

The next step is to lay out position of all joists on plate and on aisle beam, Illus. 61. Use 2 x 8 x 20', if you can buy straight ones free of knots, or 2 x 8 x 12'. If codes require 2 x 8 placed 16" on centers, do it. While not structurally necessary, you can alter position of two joists so each lines up with stall divider.

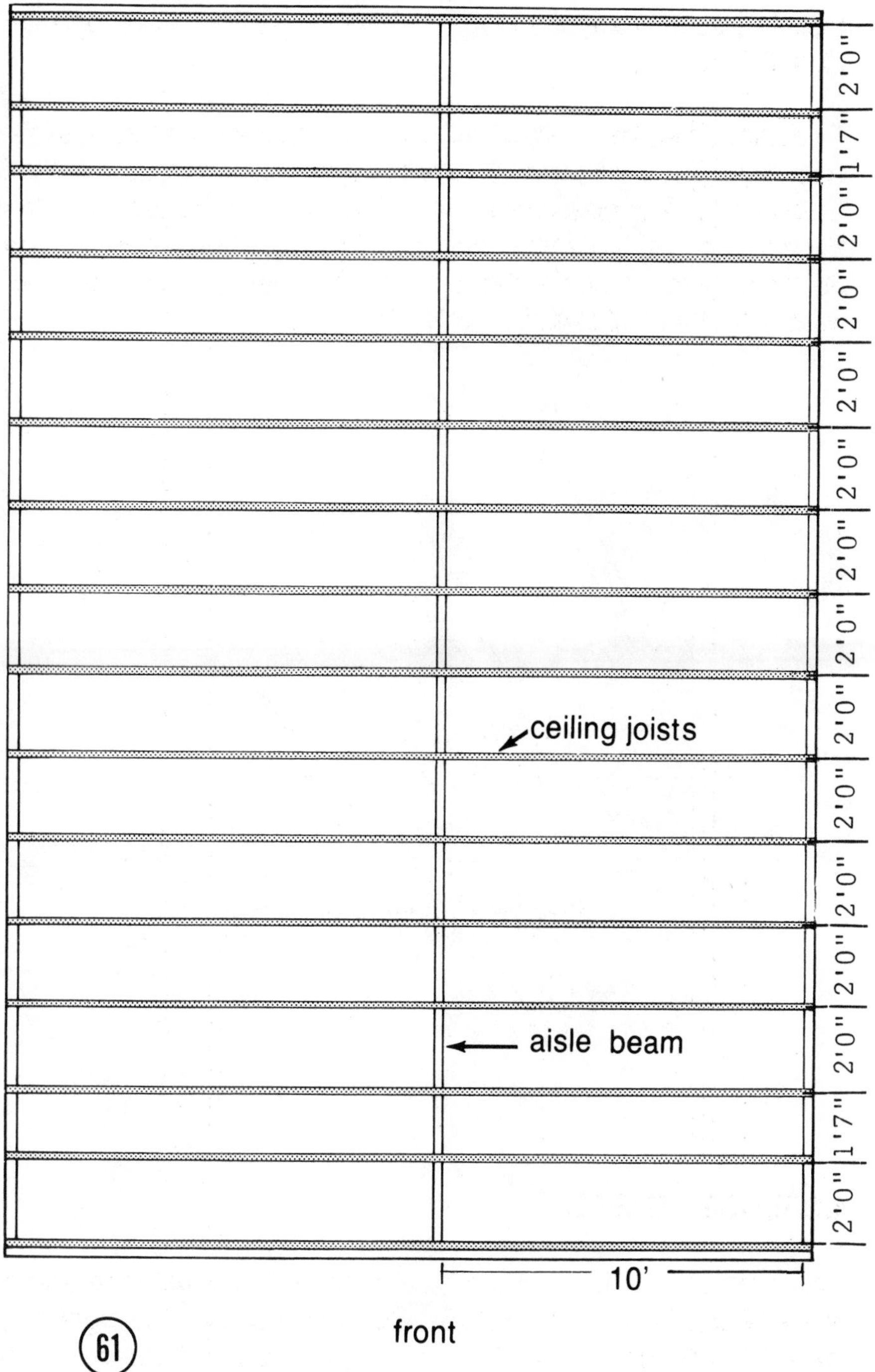

2'0"
1'7"
2'0"
2'0"
2'0"
2'0"
2'0"
2'0"
2'0"
2'0"
2'0"
2'0"
2'0"
1'7"
2'0"
ceiling joists
aisle beam
10'
front
61

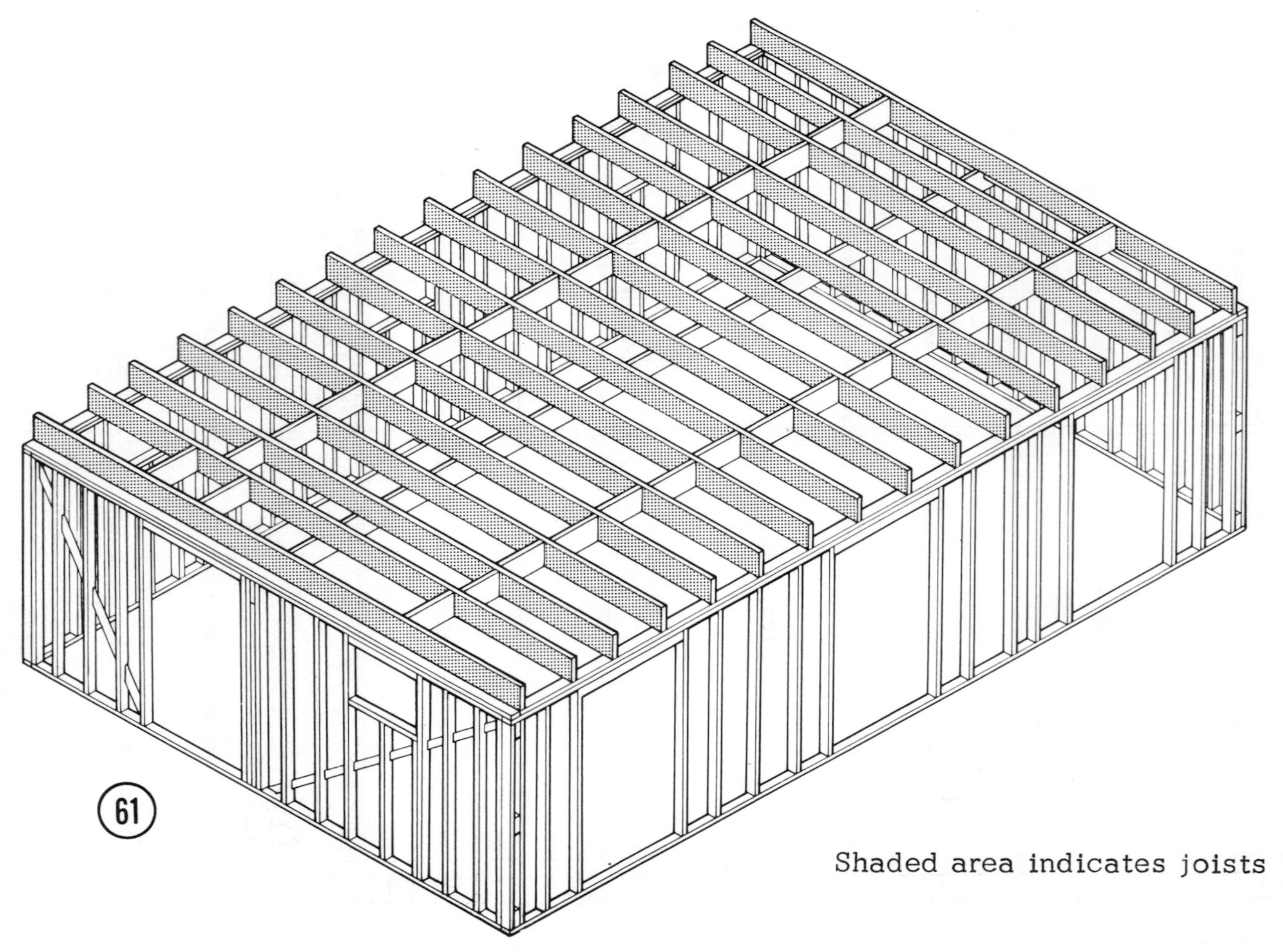

Shaded area indicates joists

The first joist is nailed 3½" in from gable ends. This places it alongside the inside edge of the plate, Illus. 62. Always sight down a joist and nail crowned edge up. To permit securing end rafter to joist, nail a filler block to gable end rafter, Illus. 82.

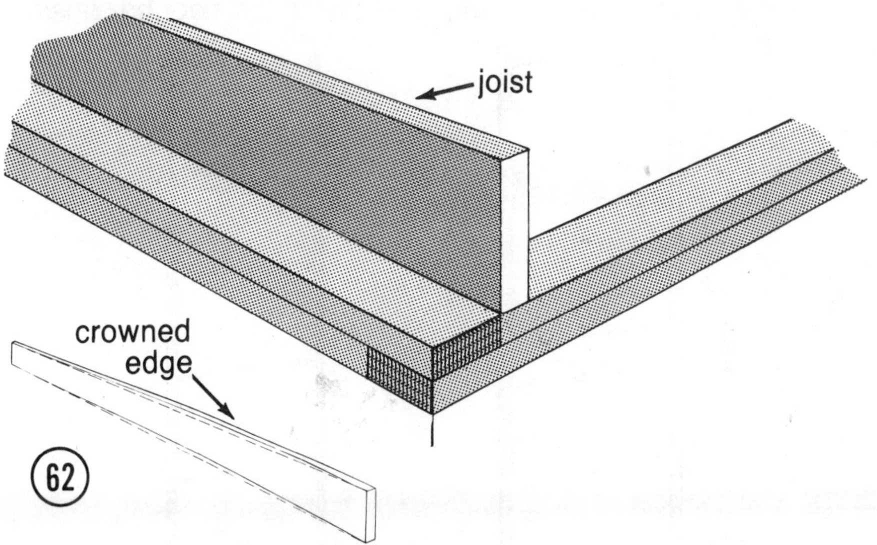

While 2 x 8's, two feet on centers, over a 10'0" span, provides ample strength; if local codes require 2 x 10, use them. If you can't obtain straight 2 x 8 x 20', use 2 x 8 x 12, overlap ends as indicated, Illus. 63. Toenail each joist to plate and beam using 16 penny nails.

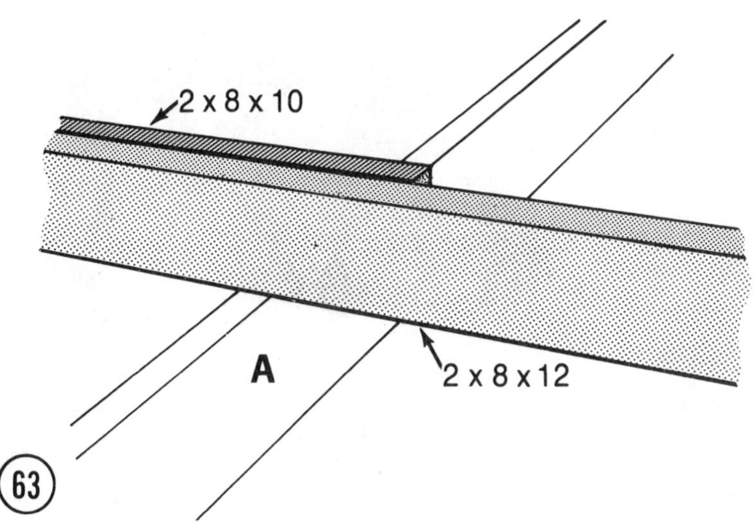

To simplify nailing joists, rent a scaffold, Illus. 64, or build one, Illus. 65, you can take apart and reassemble where needed or use saw horse brackets. Cut 2 x 4 legs to length required. Spike 2 x 8 planks for a platform. A scafford nailed to sides of building at window height, Illus. 66 simplifies laying loft flooring, raising rafters, etc.

64

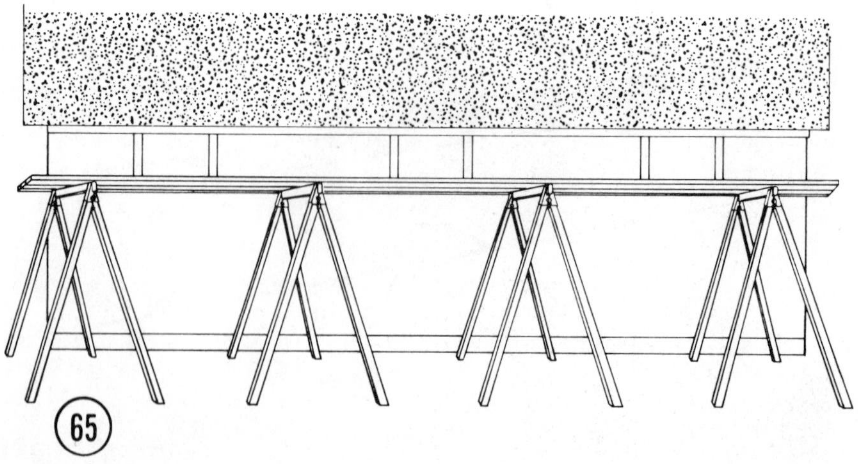

65

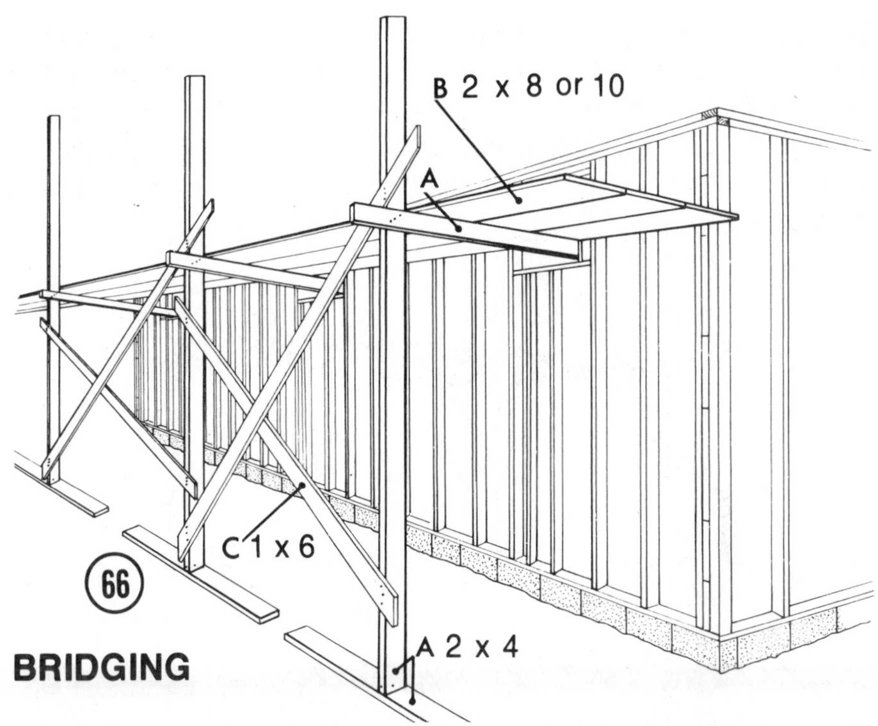

BRIDGING

After all joists have been nailed to plates and aisle beam in position indicated, Illus. 61, cut and nail bridging. This can be steel cross bridging, a stock item in lumber yards, 2 x 3, or solid bridging, Illus. 67. Use a 2 x 8, if 2 x 8 joists are being installed, 2 x 10 with 2 x 10. Nail 2 x 8 A in position indicated to provide an opening in loft. If you want to install hay racks in each stall, and fill them from the loft, nail an extra 2 x 8 B in position shown to allow for a 2' x 2' door.

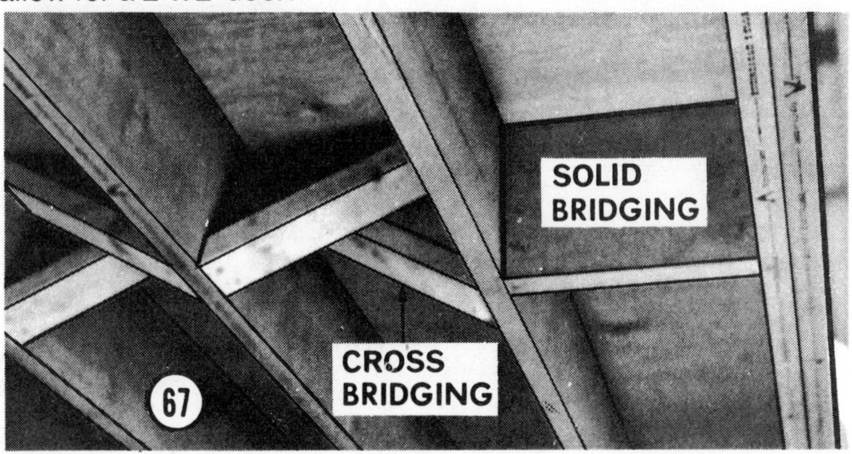

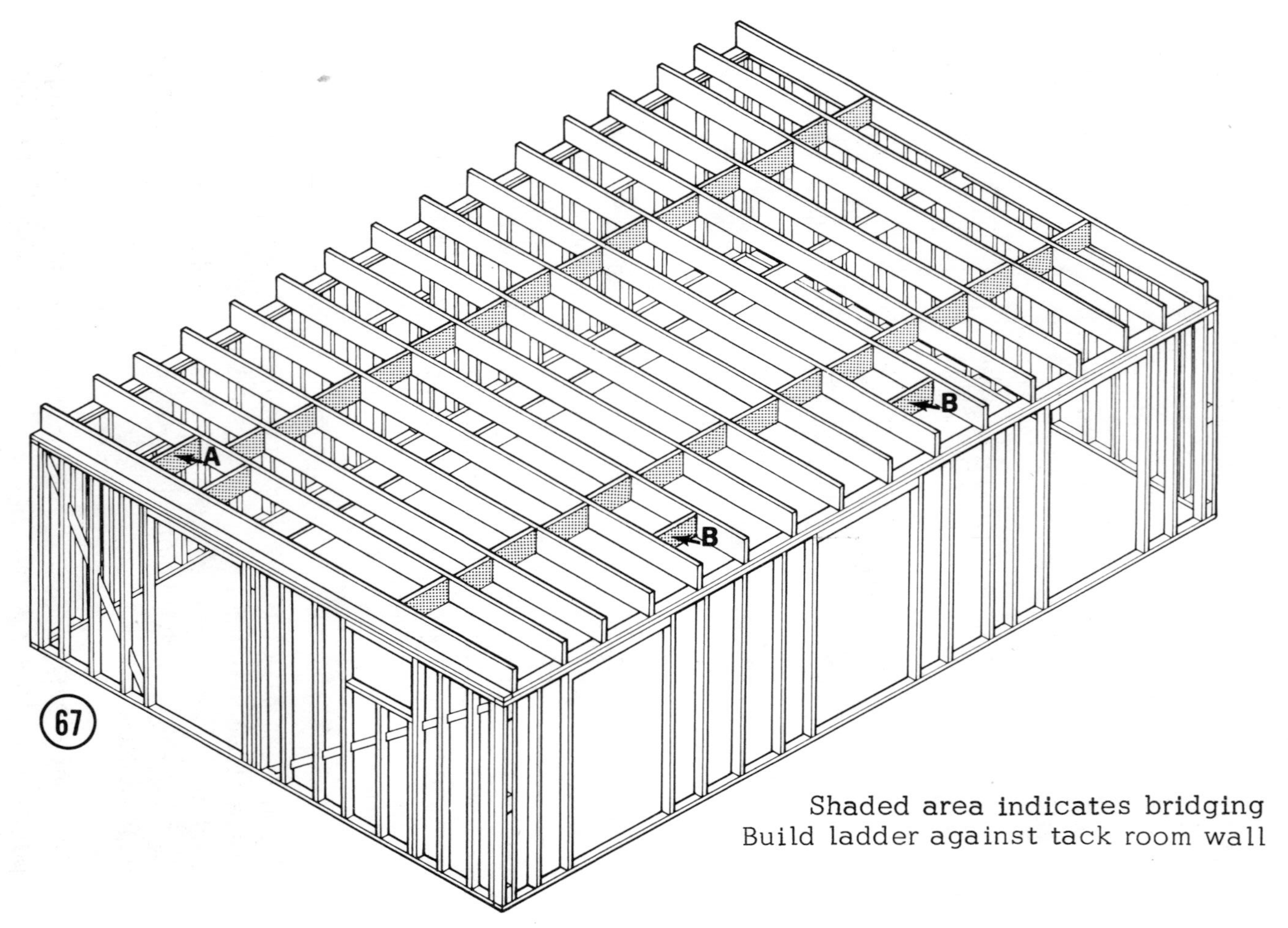

Shaded area indicates bridging
Build ladder against tack room wall

APPLY SHEATHING

In the years before plyscord solved the sheathing problem, building codes recommended stapling #15 felt horizontally to studs, Illus. 68. This helps seal exterior walls.

If felt is to be applied, start at bottom and allow first horizontal course to overlap foundation about 2". Overlap each course 2".

68

If you want to finish barn with barn battens, apply ⅝" exterior grade plywood rather than plyscord. Apply panels vertically. Butt each panel together over a stud. Position panel so it's ½" lower than top edge of lower plate, Illus. 68. Cut felt off flush with edge of plywood sheathing.

Nail panels every 6" along edges, every 8" along studs using 8 penny nails. Illus. 69, indicates application schedule. When side walls have been sheathed up to 8'0" height, replace side wall scaffolds. This simplifies raising and bolting rafters, Illus. 70.

½" Sheathing on sides projects ½" * to cover edge of sheathing on front and back.

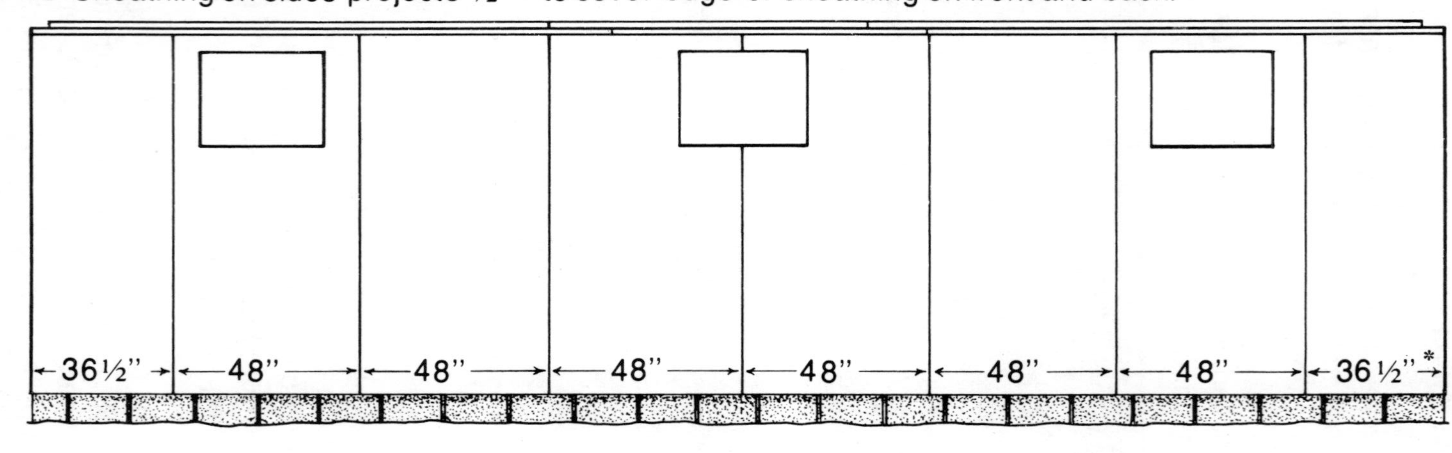

(69) * ½" for ½" plyscord
5/8" for 5/8" plyscord

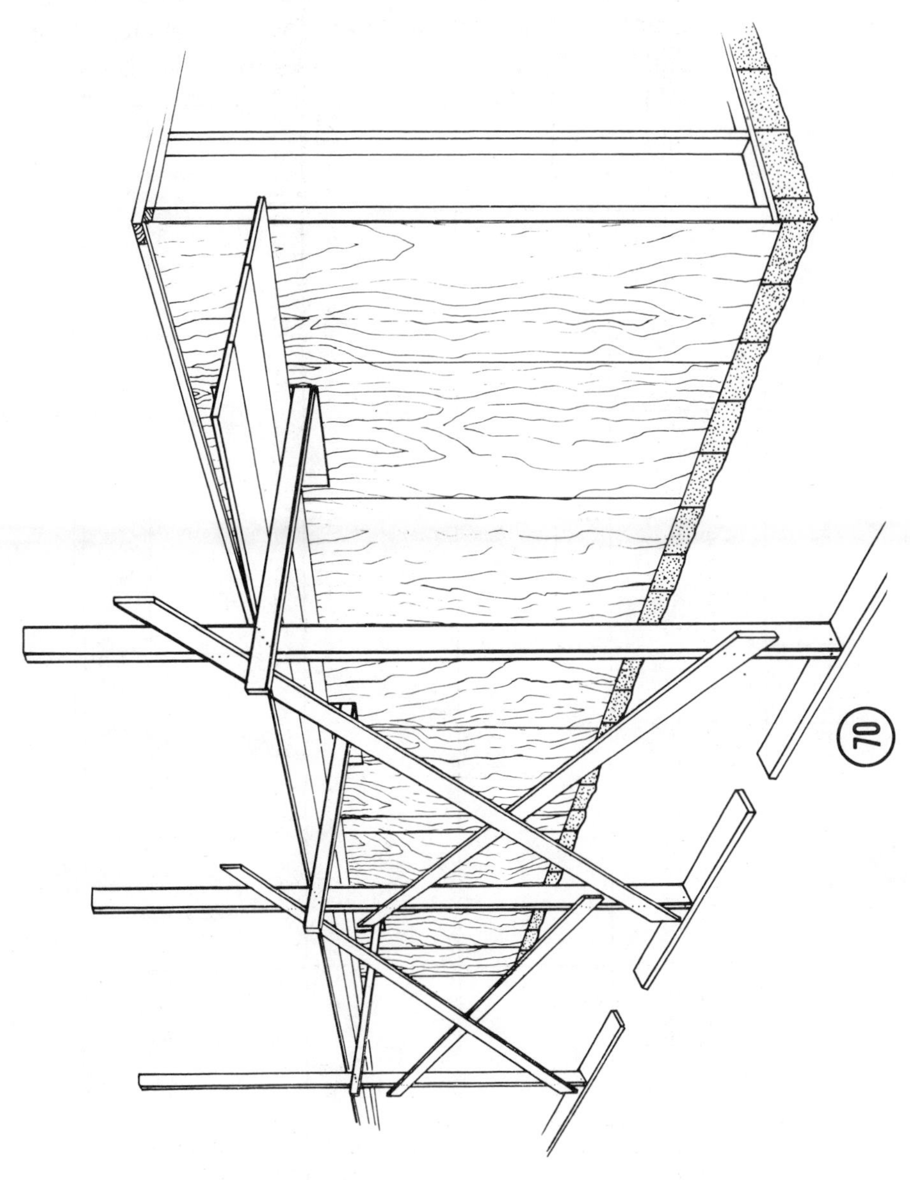
70

LOFT FLOORING

Use ⅝" or ¾" x 4 x 8 plyscord panels for loft flooring. Start at center and nail to joists every 12" using 8 penny nails. Do not nail panels on outer 2' at this time. Always butt panels together over center of joists. Apply panels in position shown, Illus. 71.

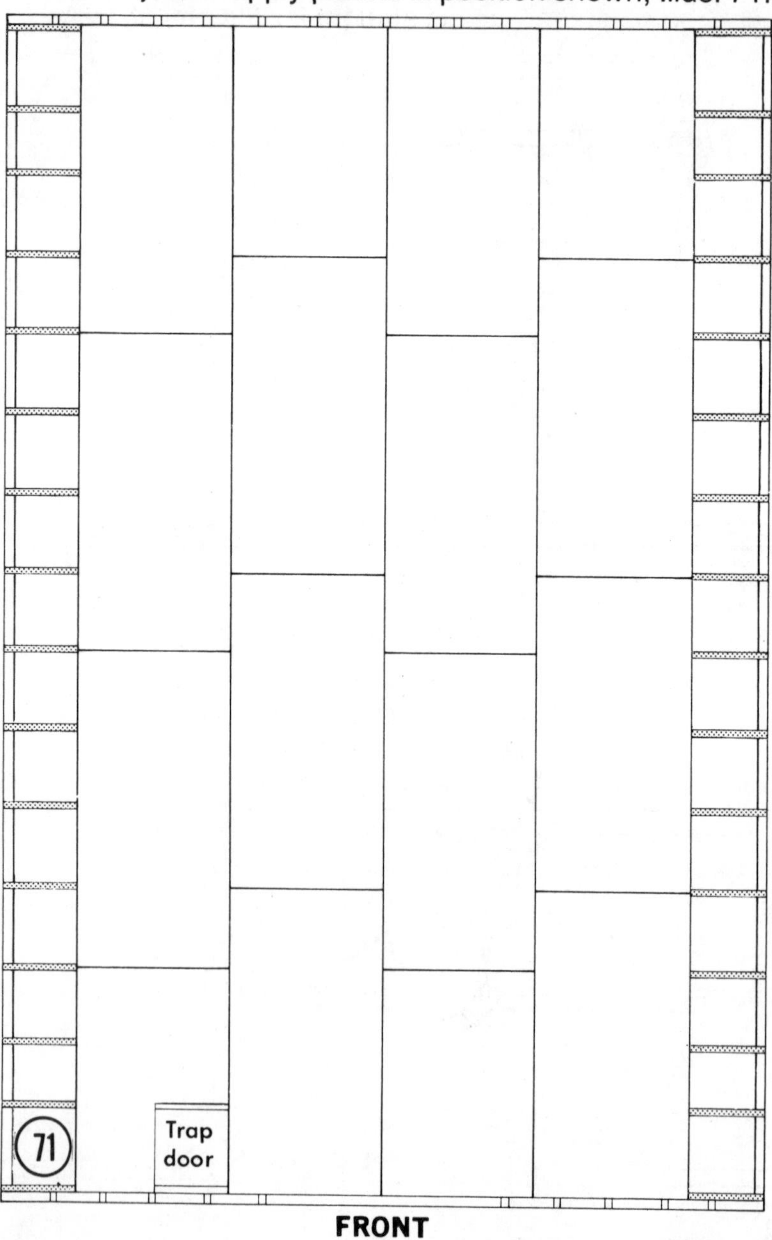

RAFTERS

When plyscord is nailed to joists, you have a platform that permits laying out, assembling and raising gambrel rafters, Illus. 72. Many lumber retailers prefabricate rafters. If you prefer making them yourself, select straight 2 x 6's free of knots.

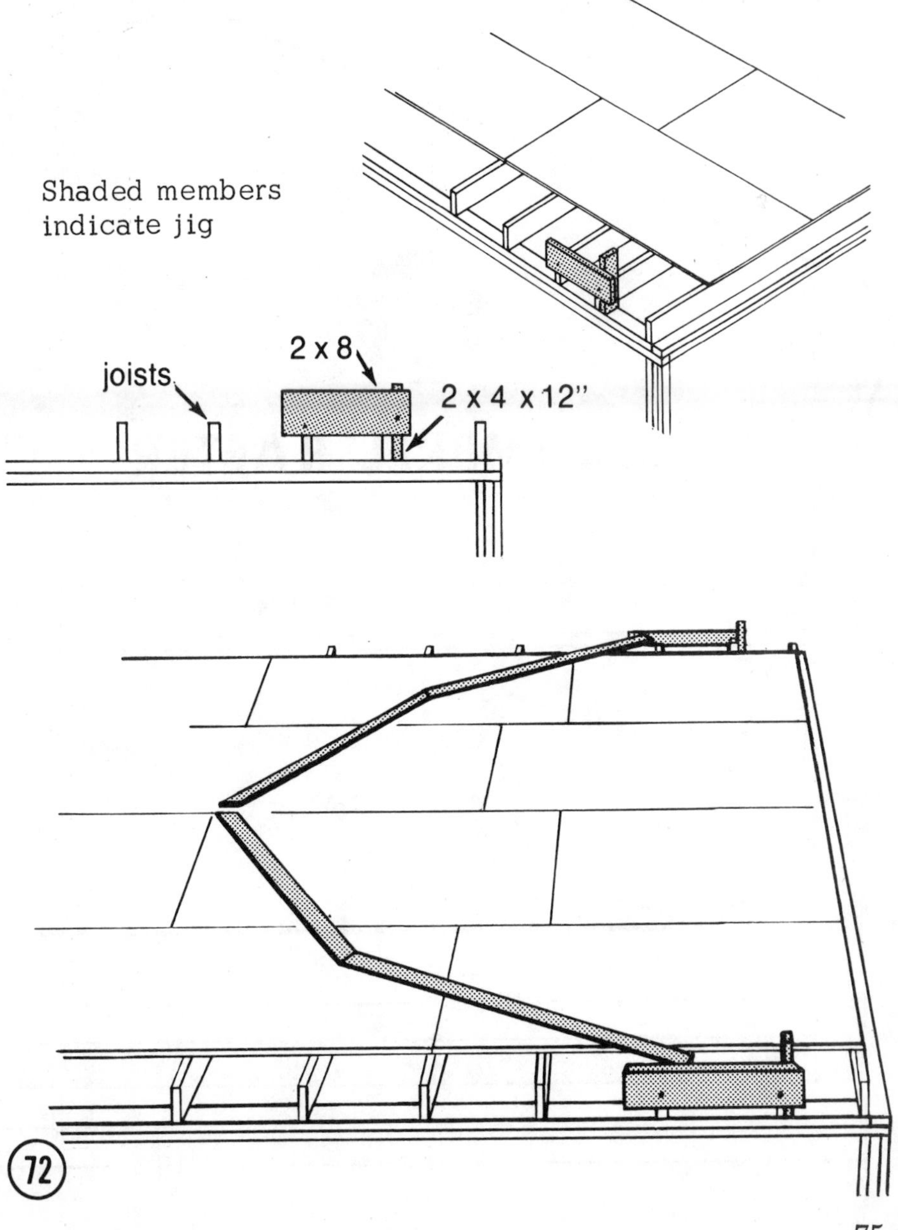

FULL SIZE ANGLE ENDS

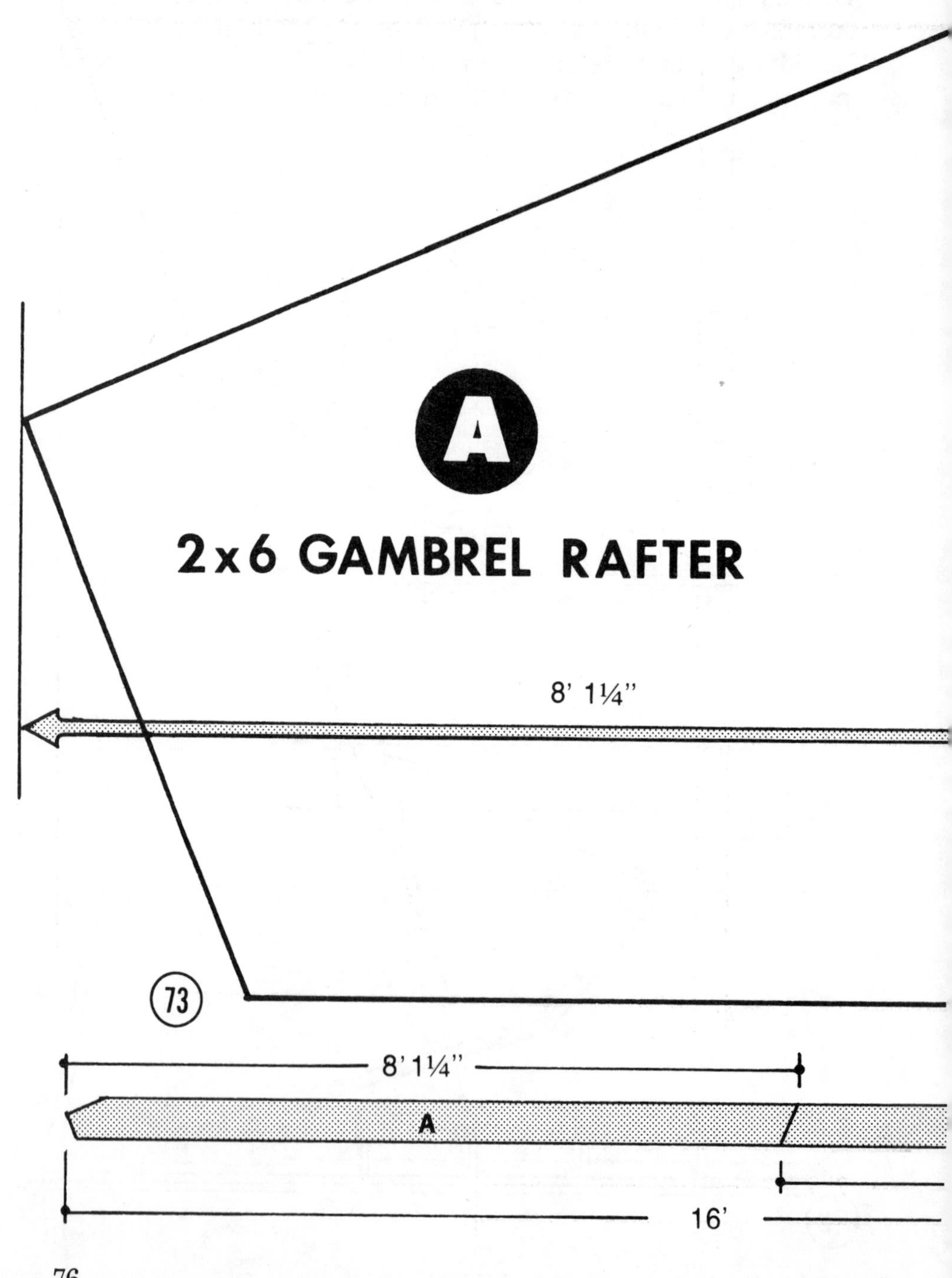

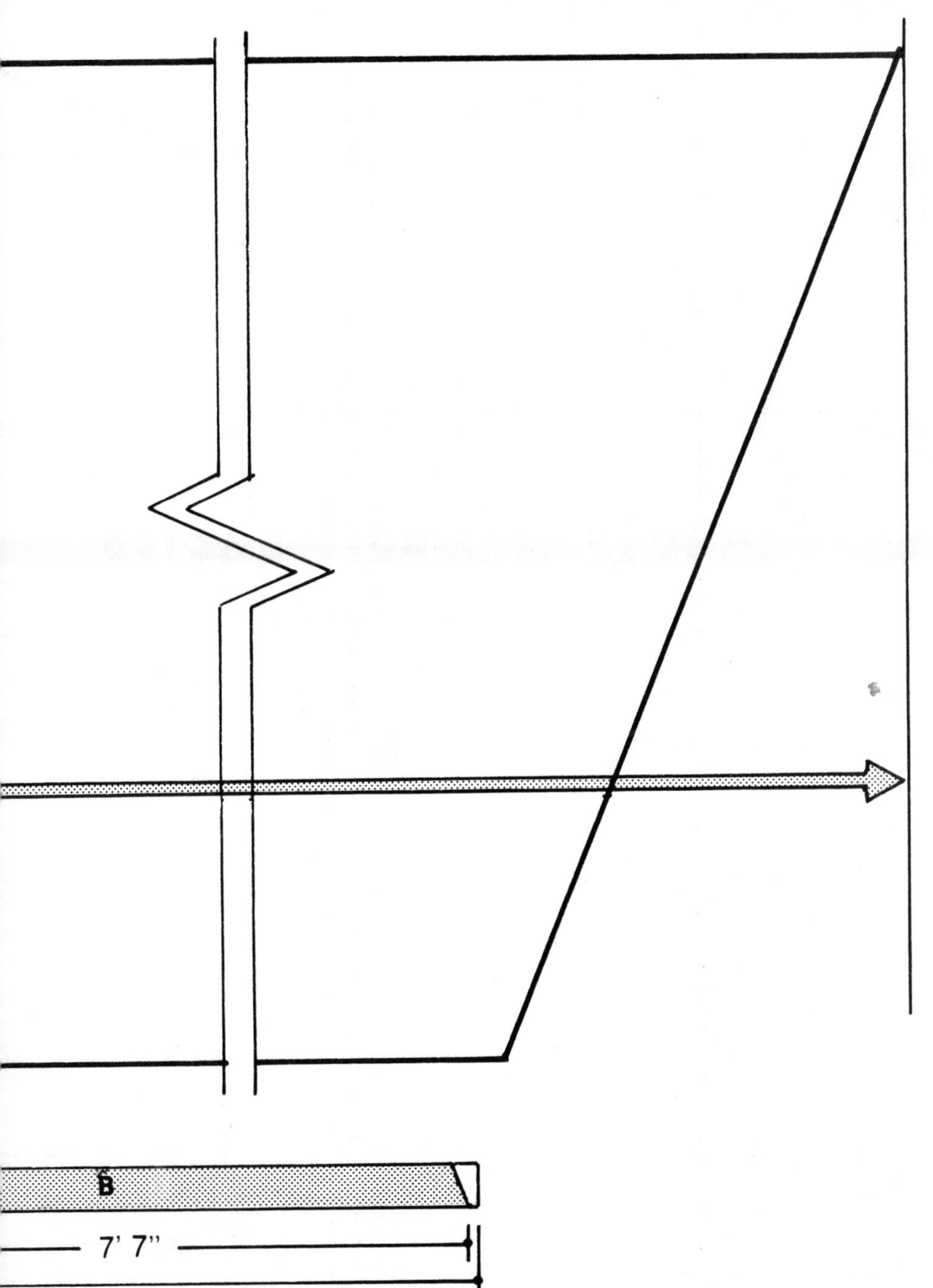
B
7' 7"

FULL SIZE ANGLE ENDS

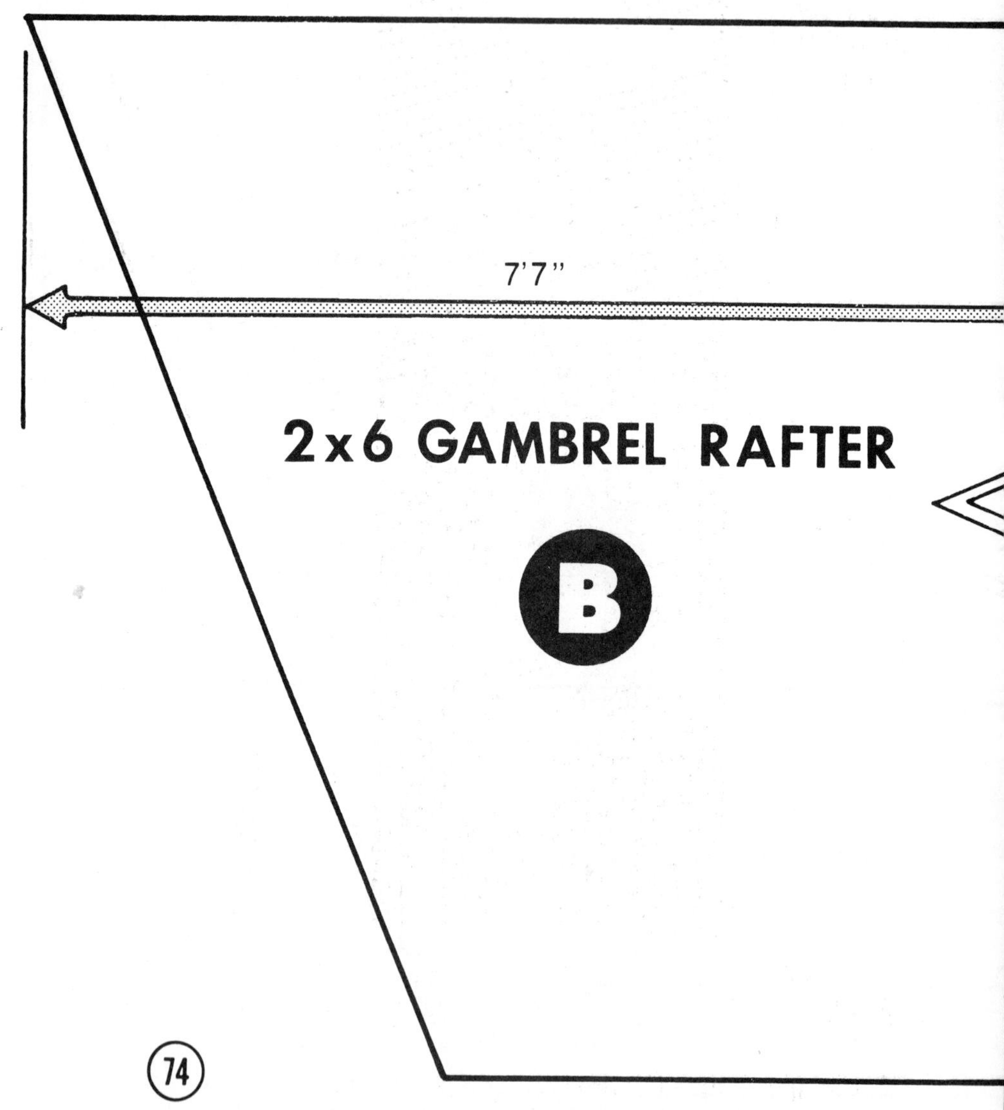

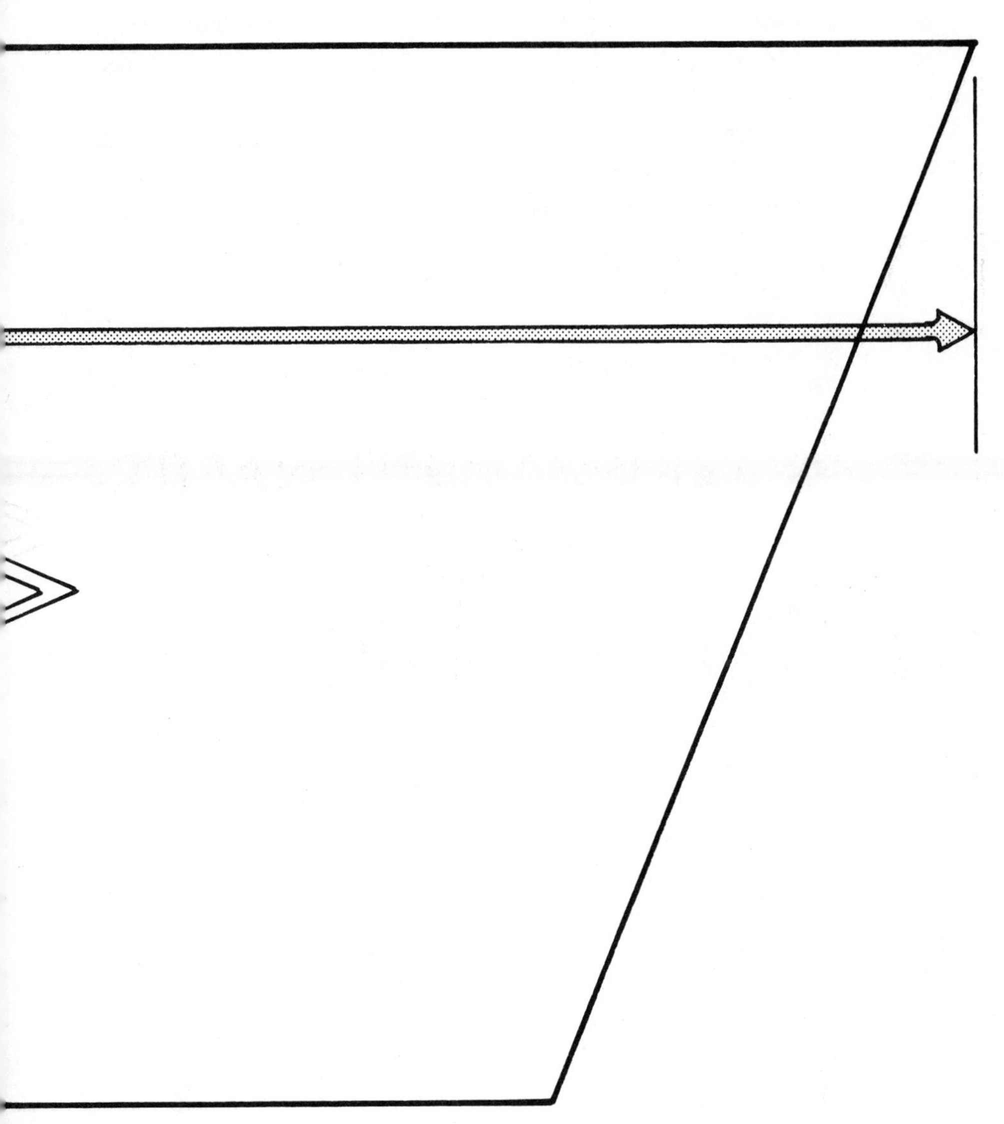

The rafters recommended contain 2 x 6 A, Illus. 73; 2 x 6 B, Illus. 74; 2 x 4 x 6'0" brace C, Illus. 75; and 2 x 4 x 8'0" collar beam D, Illus. 76; plus either wood gussets or panel clips on both sides of joint between A and B. Place A and B in position and toenail A to B with an 8 penny nail. Hold joint with a ¾" plywood or 1 x 6 gusset plate, Illus. 77, with 8 penny nails; or #10H panel clip measuring 3-7/16 x 12½", Illus. 75. Panel clips are easy to apply. Just center them over joint and pound in with a 20 oz. hammer. Only apply a wood gusset plate on inside face of end rafters, on both faces of all other rafters. If panel clips are used instead of a wood gusset plate, these can be applied to both sides of all rafters including end rafters. Next nail 2 x 4 x 6' brace C in position noted to both sides.

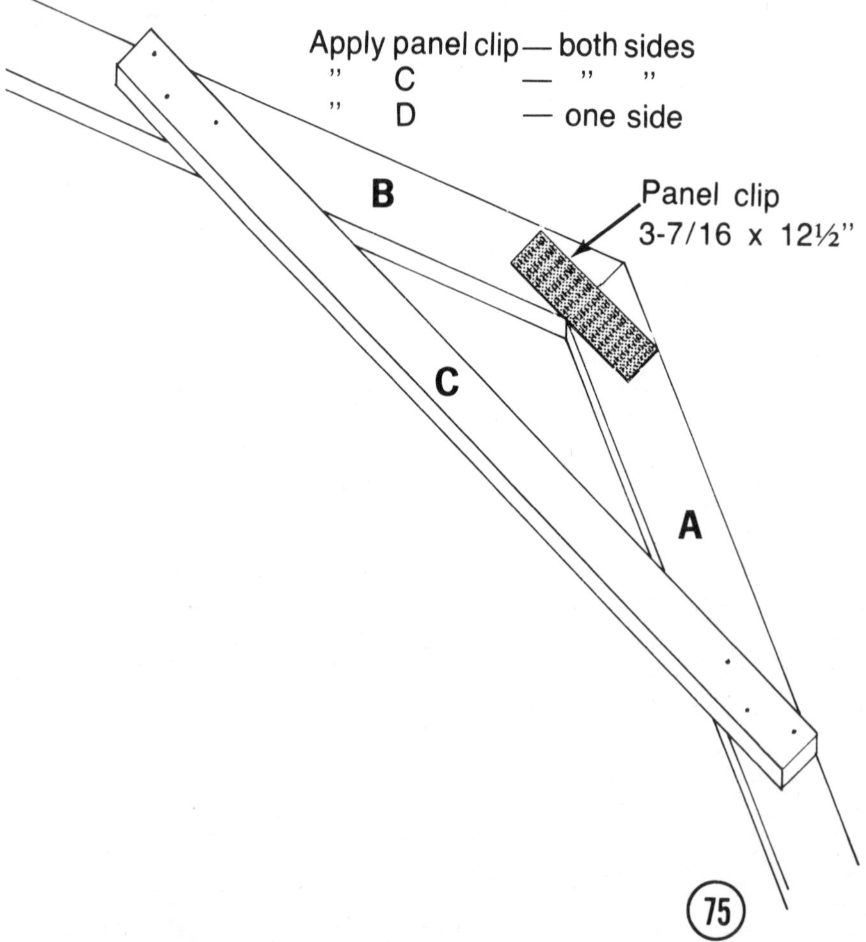

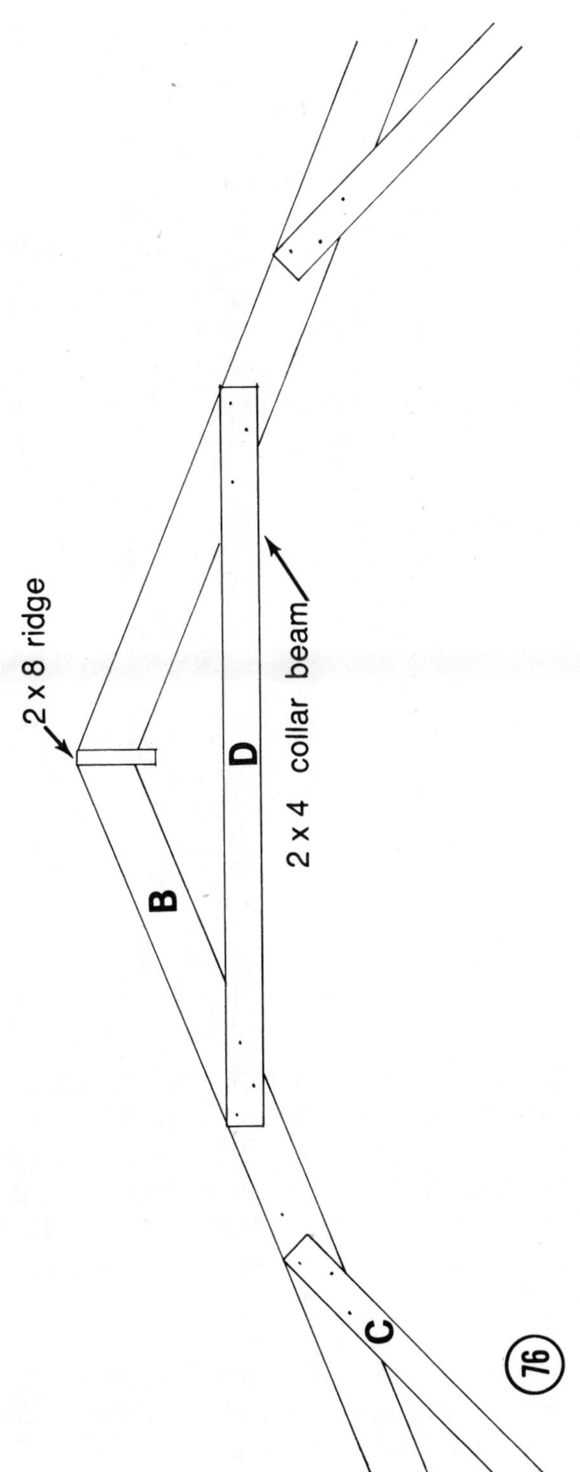
2 x 8 ridge
B
D
2 x 4 collar beam
C
76

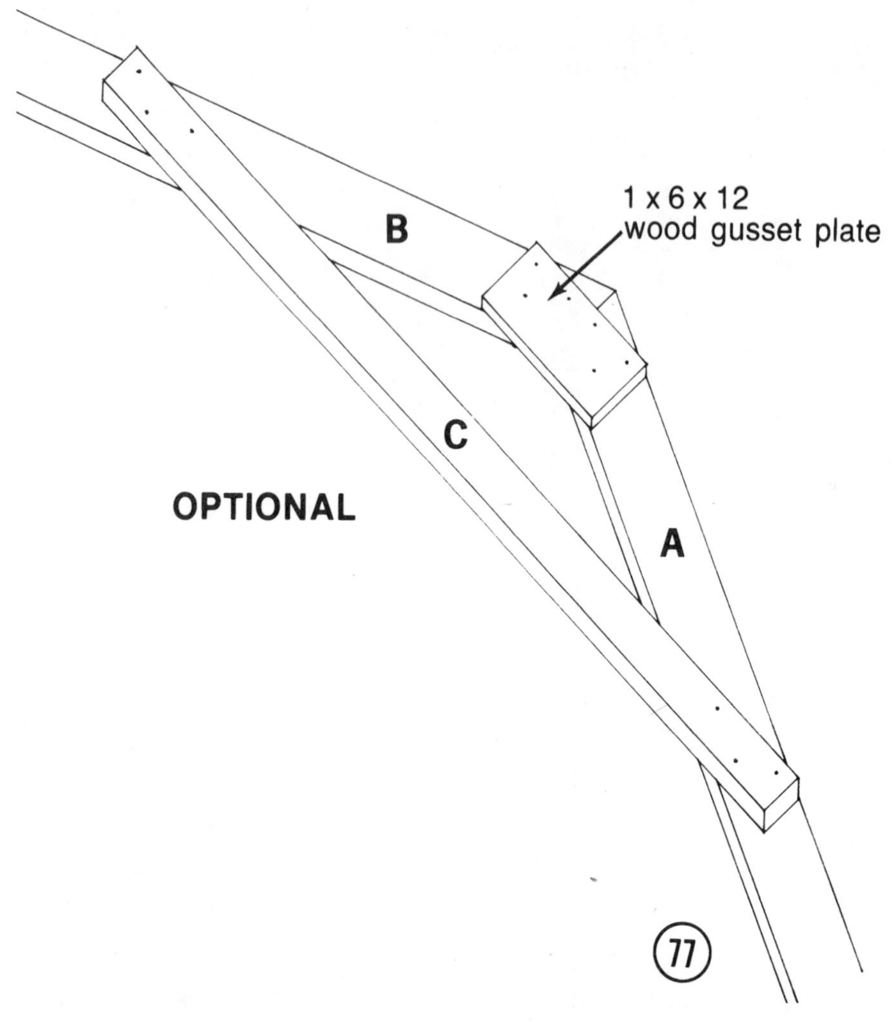

To compensate for any variation in construction, cut and test one pair of rafters using 1 x 6 and a scrap piece of 2 x 8 for ridge. Test the pair at ends, and again every 10'0". If test pair stands squarely on plate, and butts squarely against ridge, cut 2 x 6 to same size. If they don't test O.K., cut another pair to length required. When a pair proves O.K., use it as a pattern to cut others.

When you have one pair of rafters that are O.K., make up a jig, Illus. 72, 78. Nail 2 x 4's in position indicated. Nail a 2 x 8 and a 2 x 4 stop to end of floor joists in position noted.

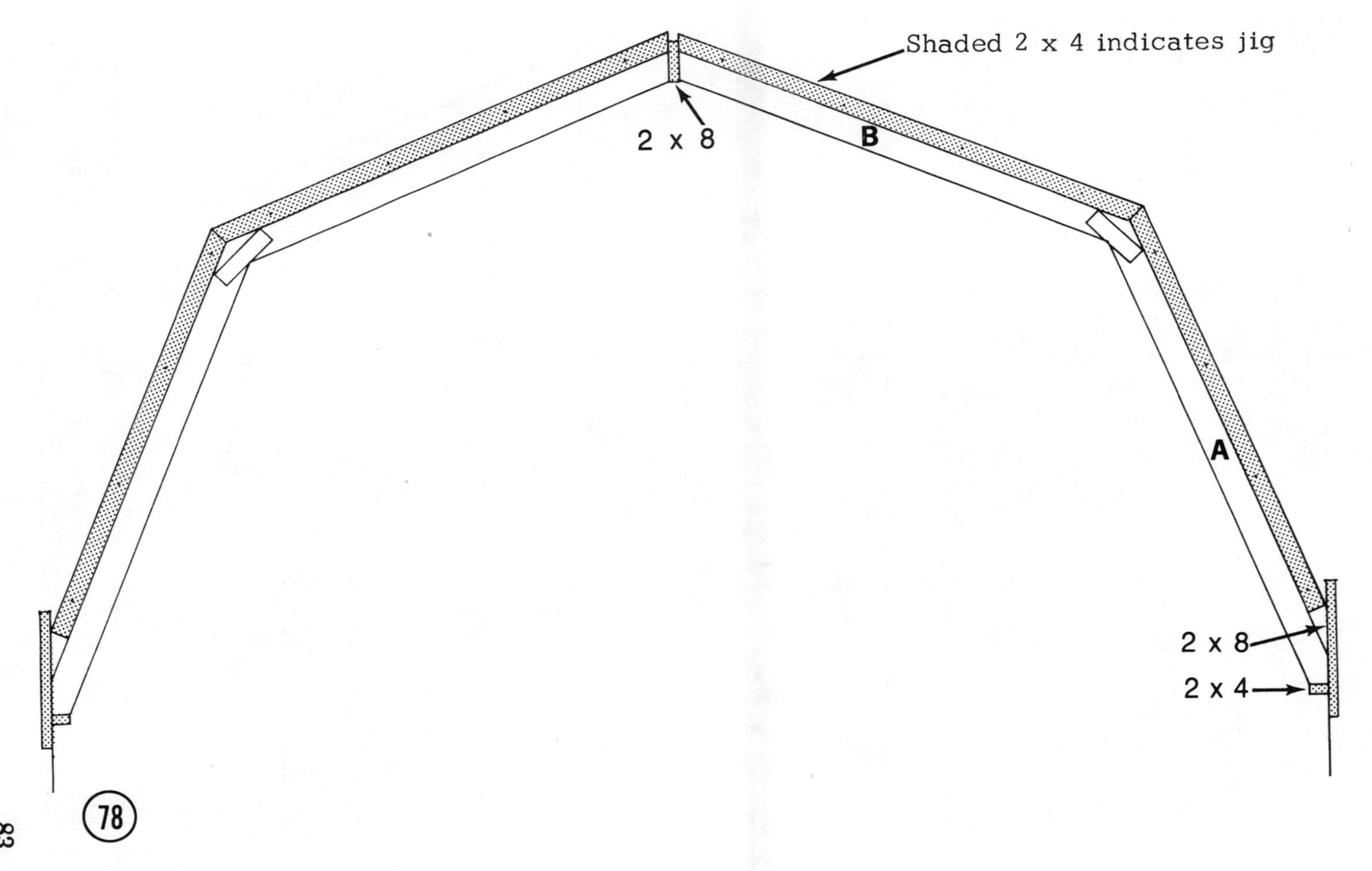
Shaded 2 x 4 indicates jig
2 x 8
B
A
2 x 8
2 x 4
78

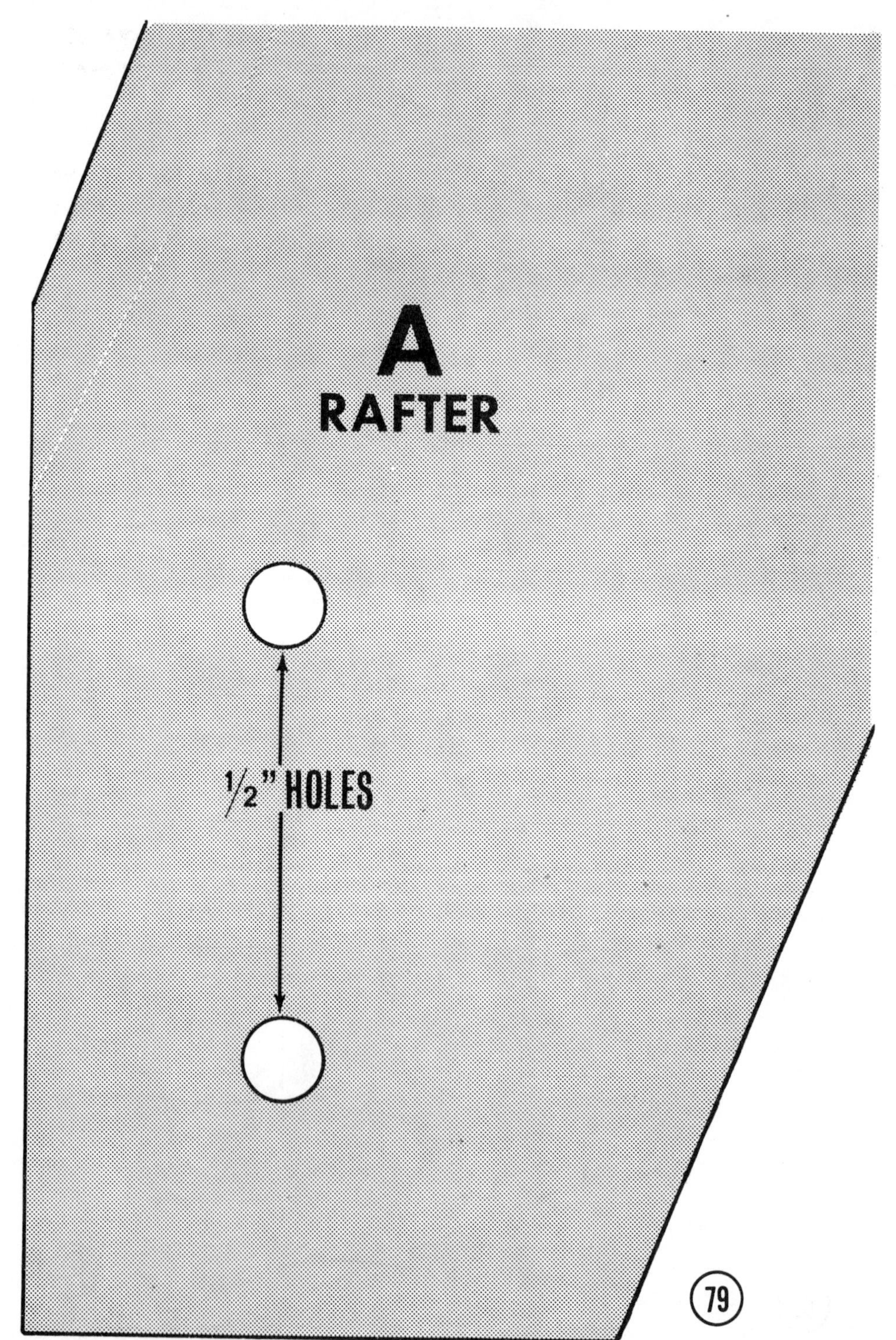
A
RAFTER
½" HOLES
79

Cut all A and B to angle and length your construction requires. Drill two ½" holes in A in position indicated, Illus. 79. Mark each assembled AB either L for left, or R for right side.

The 2 x 4 x 8' collar beam D, Illus. 76, is nailed in place after all rafters have been raised and after roof sheathing has been applied. Always nail collar beam on opposite side of cross brace, if only one C is used, Illus. 80.

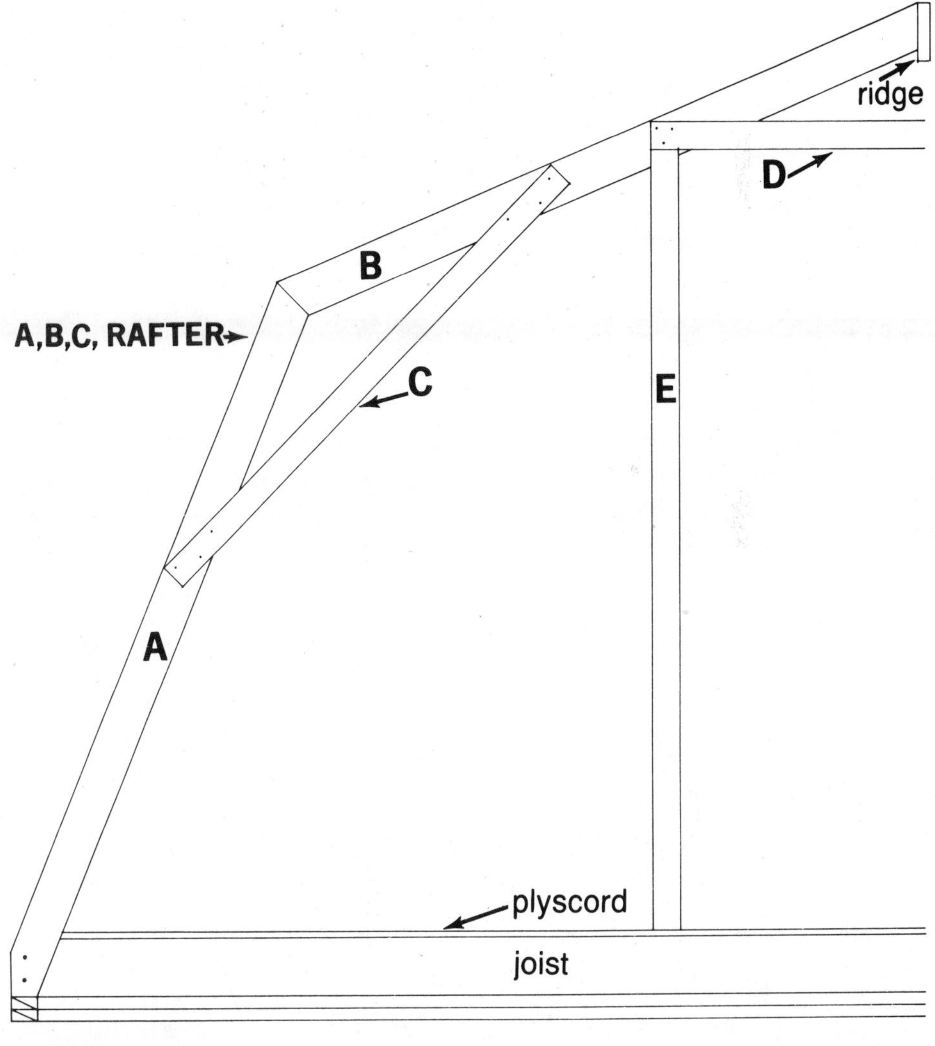

If you don't rent a pipe scaffold, or use planks on sawhorses, make up a portable scaffold, Illus. 81, using 2 x 4 for legs A, 1 x 6 for cross braces B and C. Nail B to A at a height that permits working comfortable when nailing rafters to ridge. Nail 2 x 8's to B for a platform.

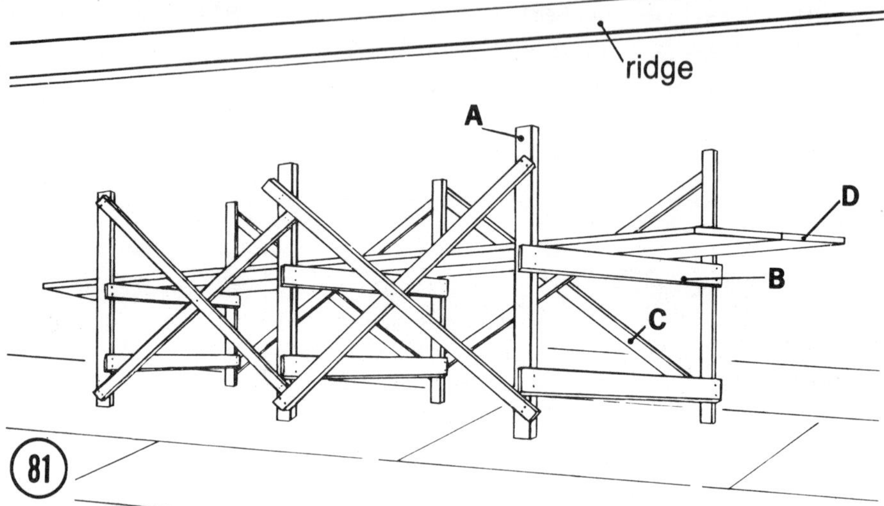

Nail 2 x 8 x 8" and ½ x 7¼ x 8" filler blocks, Illus. 82, to inside face of first rafter. Place first rafter in position flush with outside edge of plate, Illus. 83. With rafter flush with edge of plate, and with one person holding ABC in position, insert ½" drill through ½" hole in A and bore through filler blocks and joist. Fasten rafter to joist with one ½ x 6" bolt and washer, Illus. 84. Only fasten nut finger tight.

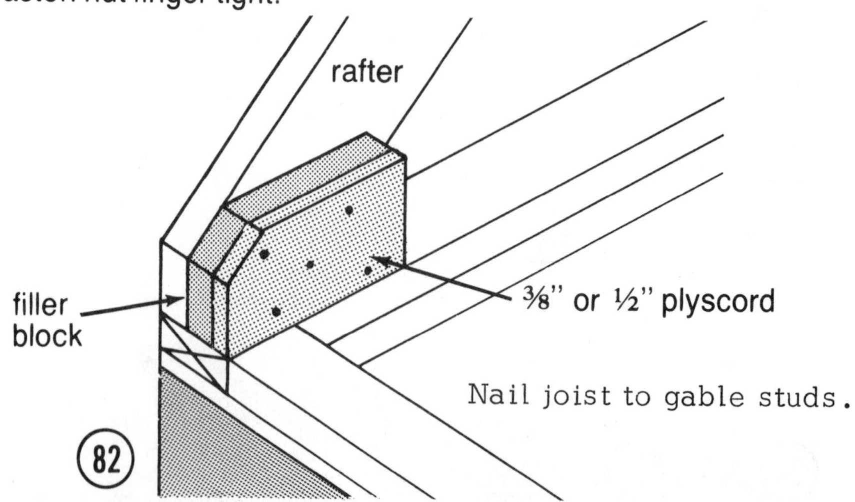

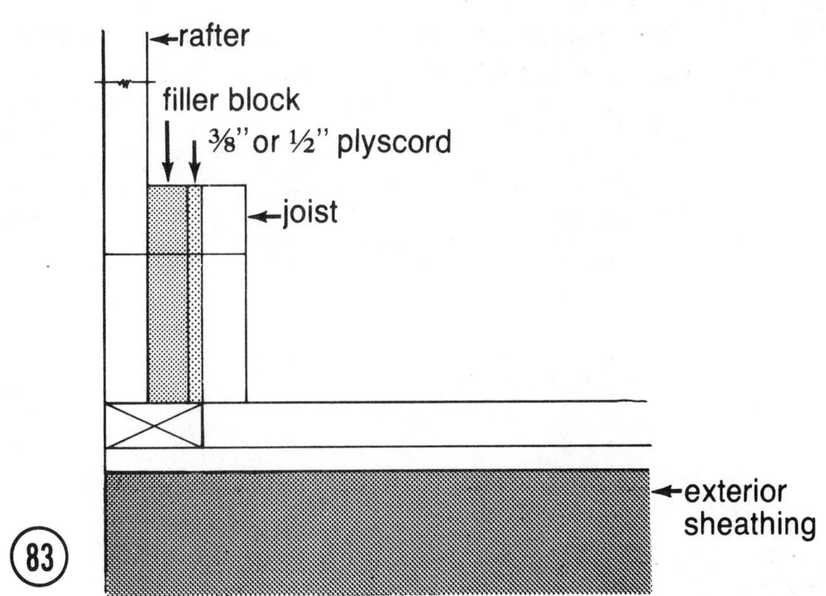
rafter
filler block
⅜" or ½" plyscord
joist
exterior sheathing
83

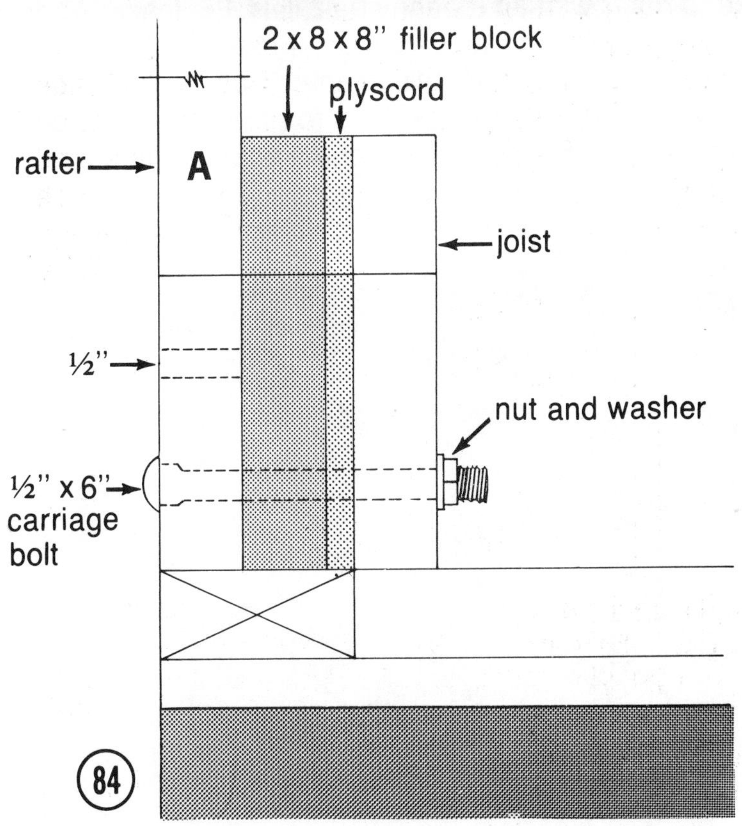
2 x 8 x 8" filler block
plyscord
rafter
A
joist
½"
nut and washer
½" x 6" carriage bolt
84

If you want the ridge to project beyond face of gable end, Illus. 85, project ridge 11¼". You will need two 2 x 8 x 16' for ridge.

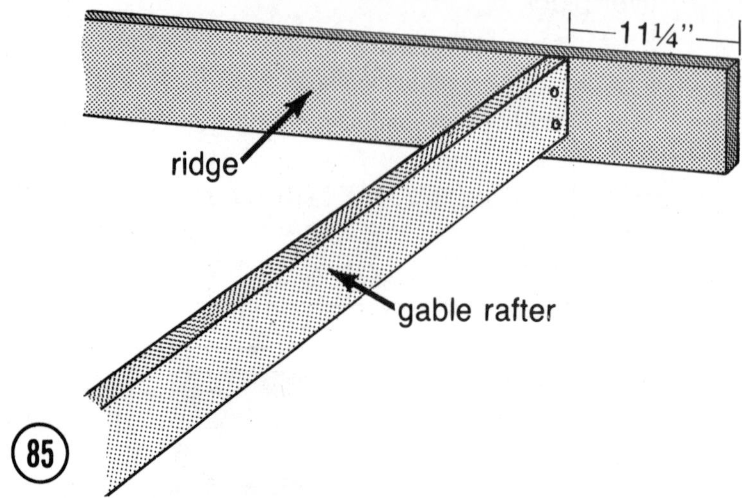

Bolt a matching rafter ABC on other side. When rafter is plumb, toenail ridge to rafter with an 8 penny nail, Illus. 86. Insert ½" bit through other ½" hole and drill through. Fasten second bolt. Spike A to joist with 16 penny nails. You can use ½ x 4" bolts and washers to fasten all rafters to joists except the end rafters. When end rafter is in position, raise the next rafter about ten feet away.

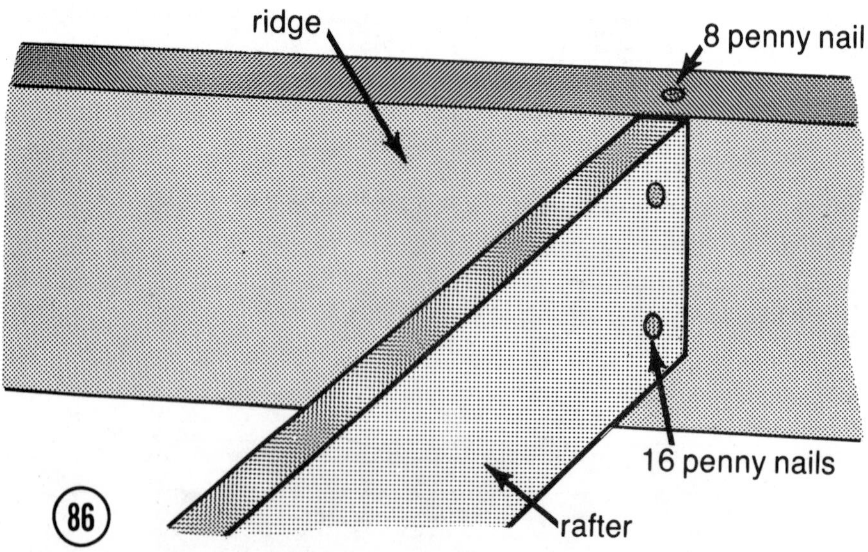

Position each rafter against ridge, then toenail through ridge into rafter with an 8 penny nail. Drive two 16 penny nails through both sides of rafter into ridge.

Check ridge with level and support free end with 2 x 4 posts placed on a 2 x 4. Raise the six rafters shown, Illus. 87. Always bolt rafter to joist, toenail ridge to rafter, then spike rafter to ridge. Drill second hole through A and fasten bolt.

After raising the six rafters, plumb gable pair by driving a nail into face of rafter just below ridge, Illus. 52. Tie a plumb line to nail. Measure distance line is from face of rafter. When point of bob measures same from face of shoe, rafter is considered plumb. Deduct thickness of sheathing when measuring line. When plumb, wrench tighten bolts to joist.

When gable end rafters and those shown in Illus. 87 have been raised and plumbed, nail a 1 x 6 diagonally across inside face to brace and hold rafters in place. Keep checking ridge with a level as you go along. Butt ends of ridge together between rafters and fasten a splice plate on one side, Illus. 56, a wood gusset plate on the other.

After all rafters have been raised, plumbed and securely nailed, bolted and braced in position, nail 1 x 6 across top of rafters to hold them exactly 2'0" apart. Chop ends of floor joists, Illus. 88 to angle of rafter.

Cut 8'0" strips of plyscord to width needed to cover top half of lower plate and end of rafter, nail in place, Illus. 96.

You can now start framing in gable studs.

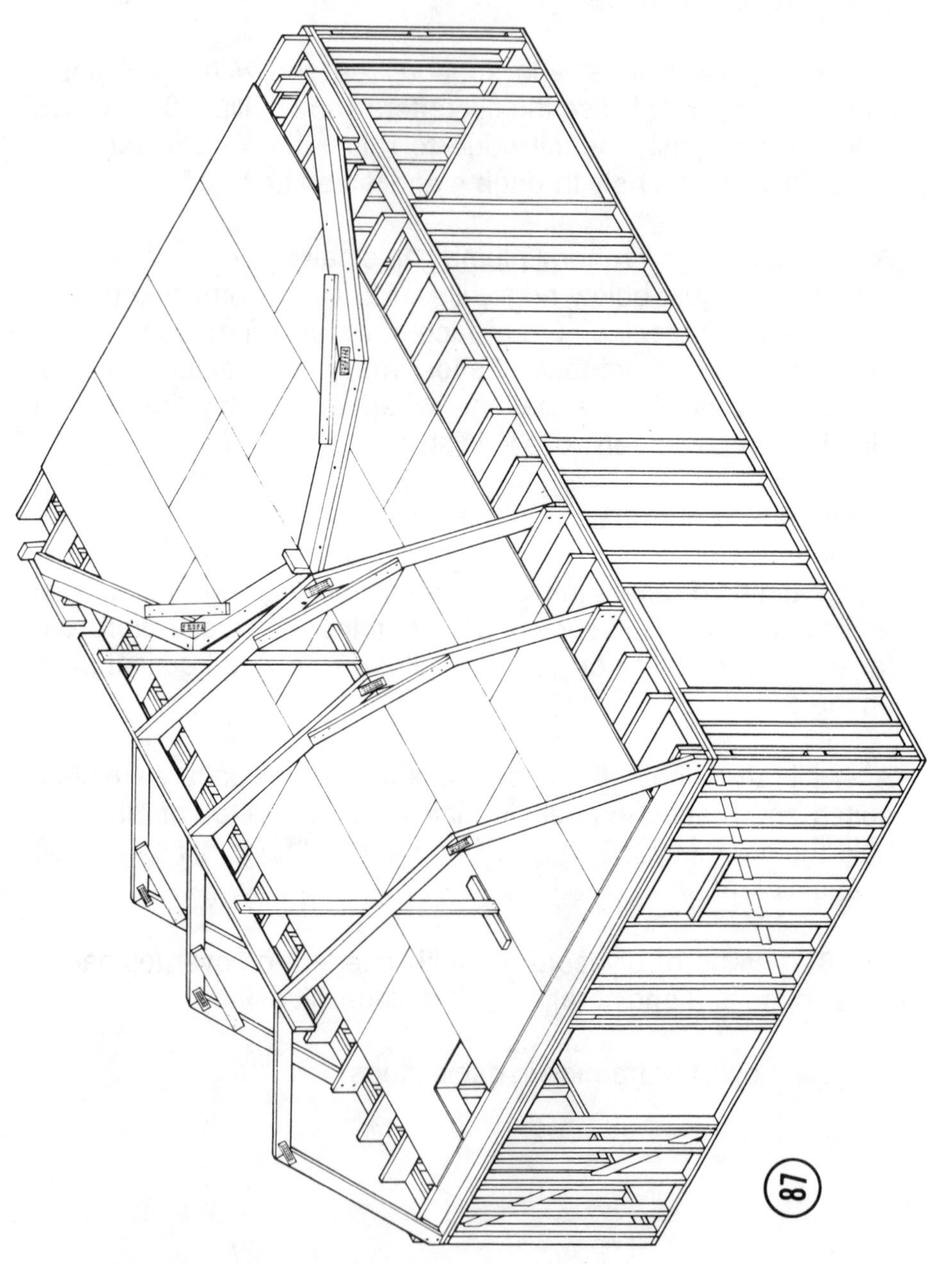
87

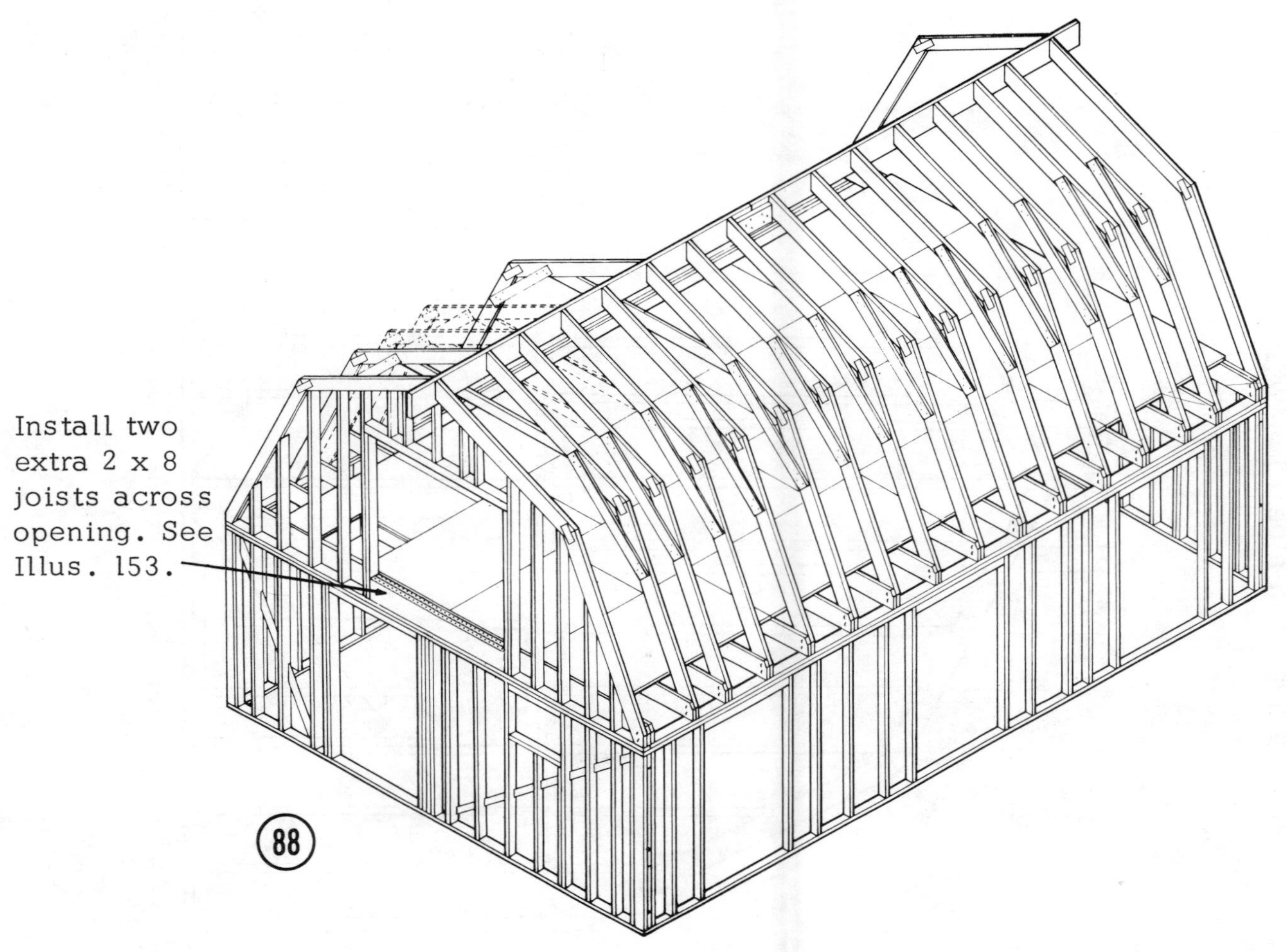
Install two
extra 2 x 8
joists across
opening. See
Illus. 153.
88

GABLE STUDS

Cut gable studs, Illus. 89, to angle and length each requires. Place a 2 x 4 in position indicated. Check with a level to make certain it's plumb. With a pencil draw exact angle and length required.

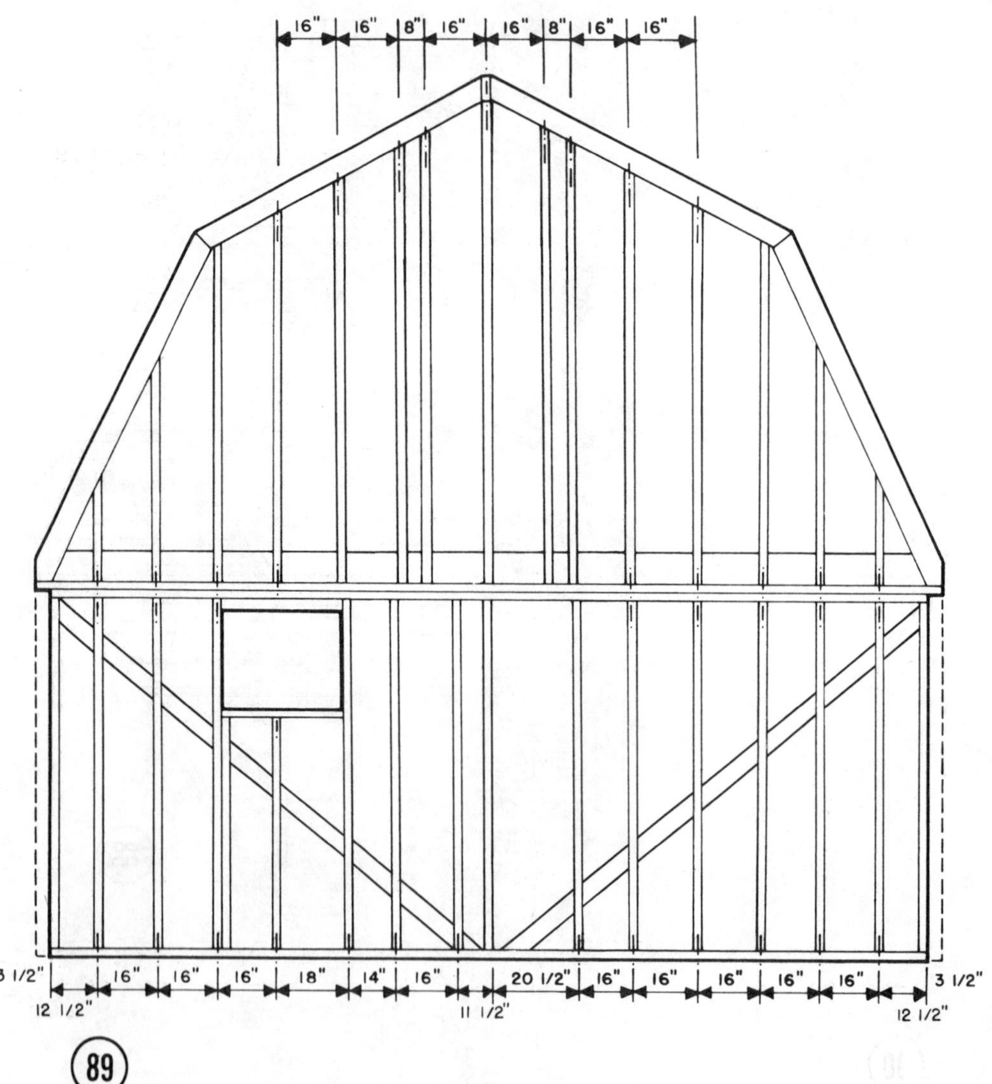

89

Cut studs to length framing for loft door opening requires. Outside door to loft can be installed at most convenient end. Nail header and sill in position indicated, Illus. 90.

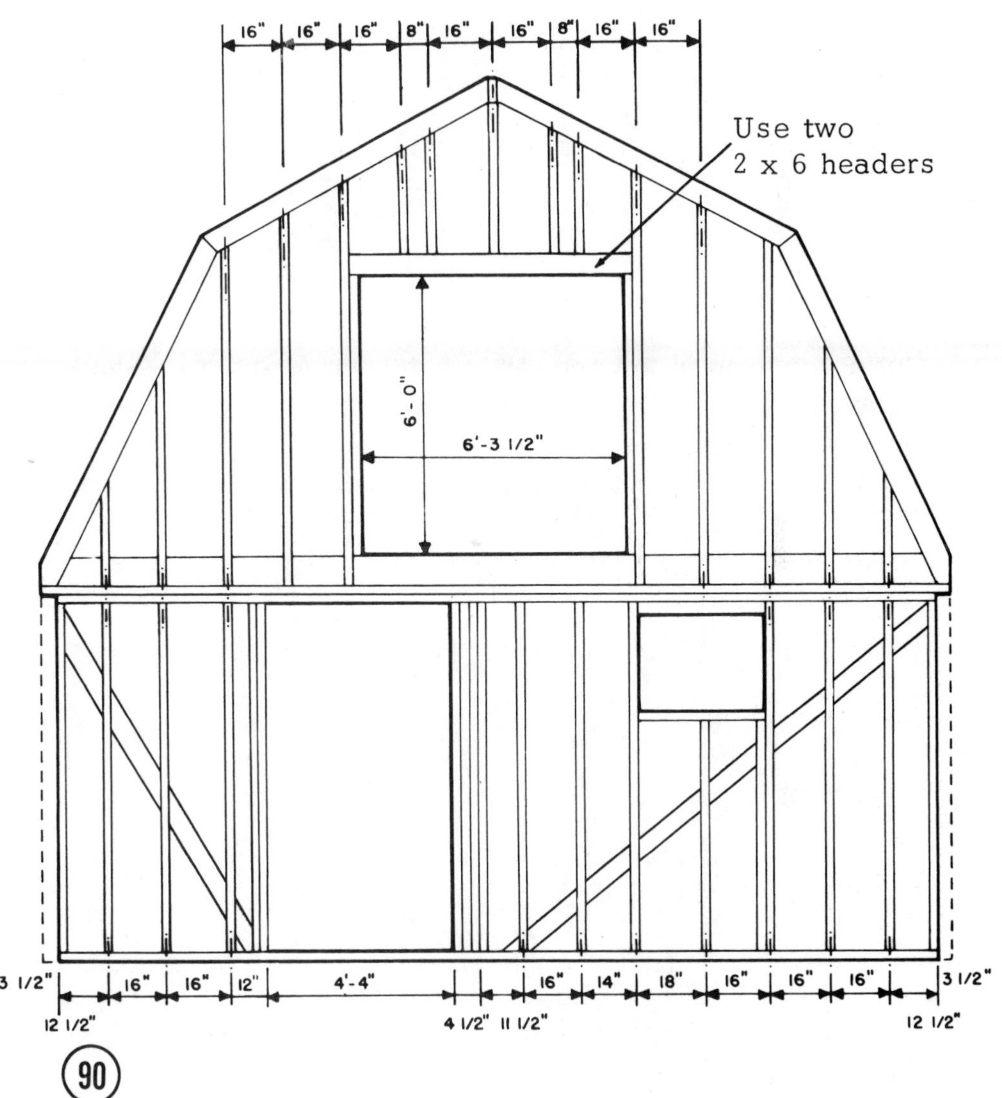

90

Build a scaffold at gable ends and apply sheathing. Sheathing finishes flush with rafters. Next finish laying floor panels. Lay ¾ x 2 x 8 panels in position butting up against rafters. With a square draw exact position of each rafter. Remove panel. Using a saber saw, Illus. 91, cut notch to depth rafter requires. Nail each panel in place using 8 penny nails.

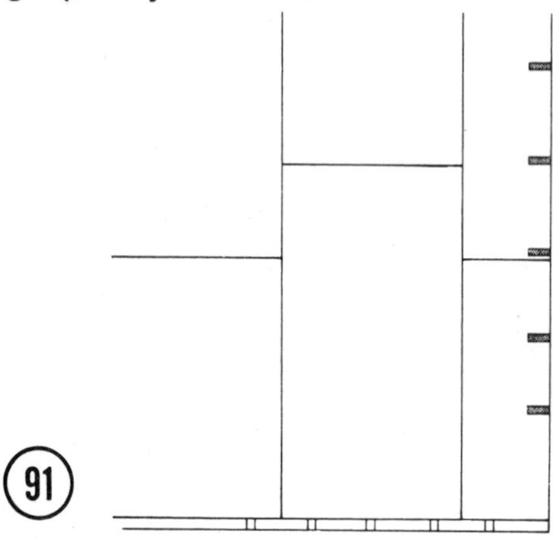

91

Cut an opening in loft floor for a ladder, Illus. 92. To make a ladder to the loft, nail 1 x 4 rungs across two studs, 12" apart. Nail 1 x 2 to edge of 2 x 4 to provide additional support between rungs, Illus. 93. Cut door panel to full width of opening. Hinge door with a pair of 3" x 3" loose pin butt hinges. Apply a door handle to top, Illus. 94.

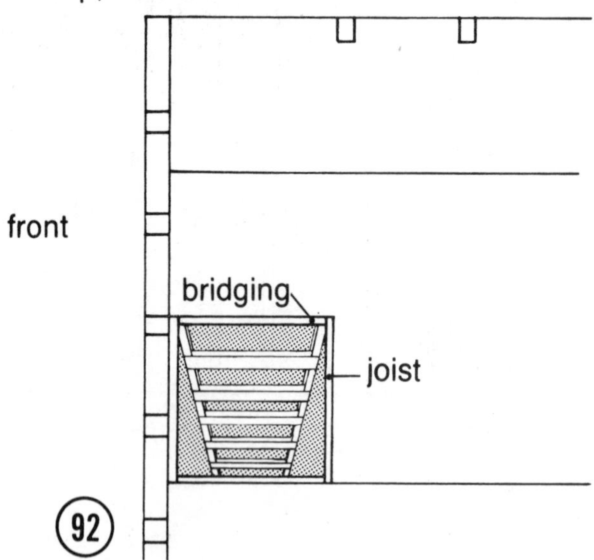

92

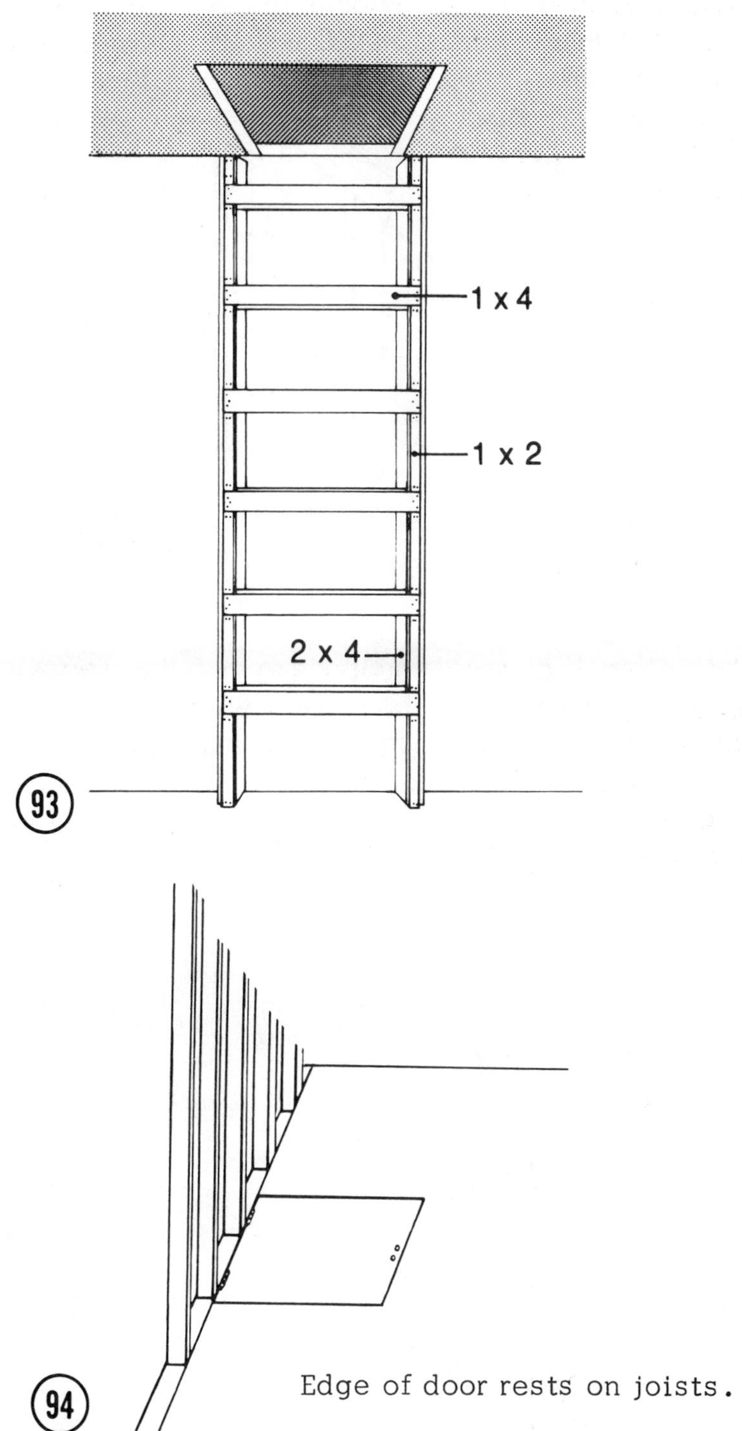

Edge of door rests on joists.

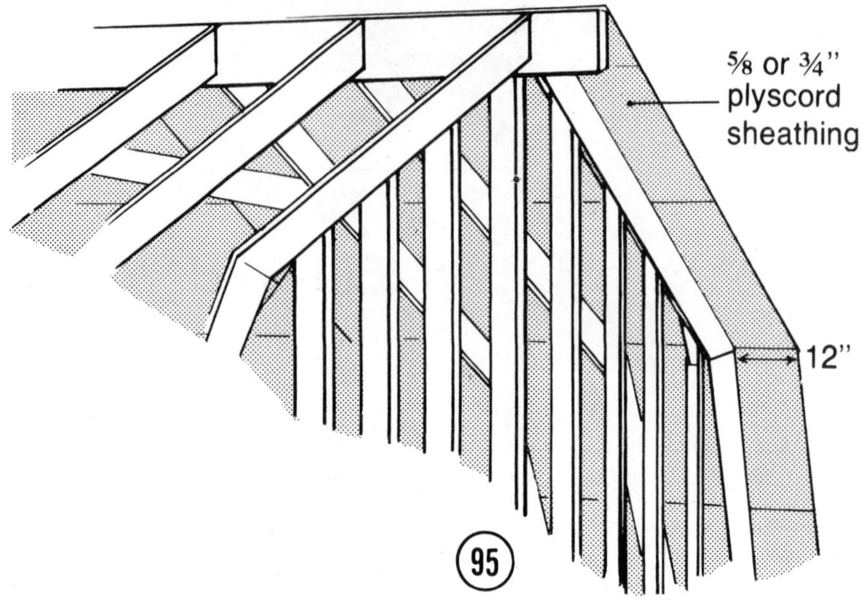

Apply ⅝ or ¾" plyscord sheathing to roof. If you want a 12" roof overhang over front and back, Illus. 95, project panel 12" over gable ends. Position first course of sheathing so it projects over sides about 1½", Illus. 96.

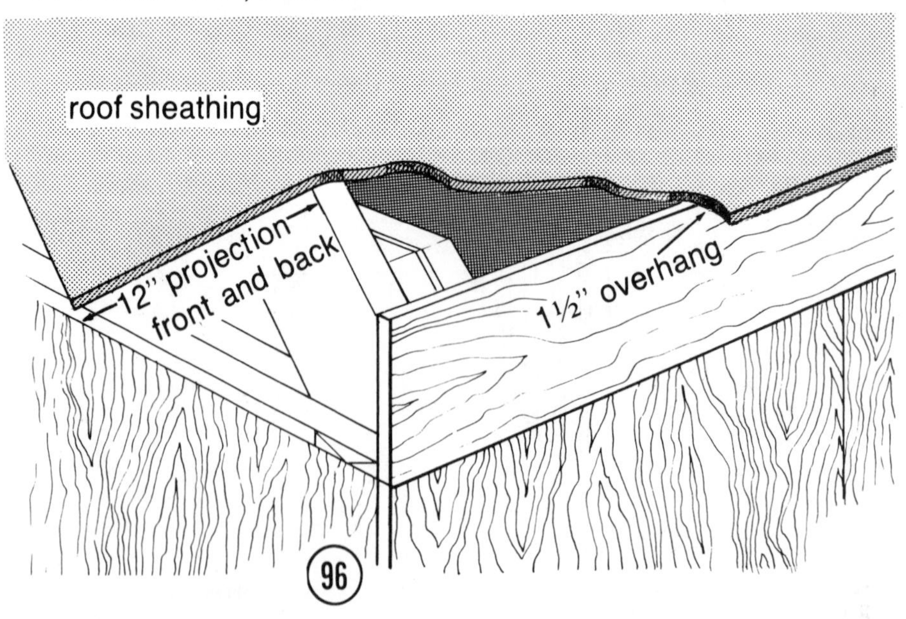

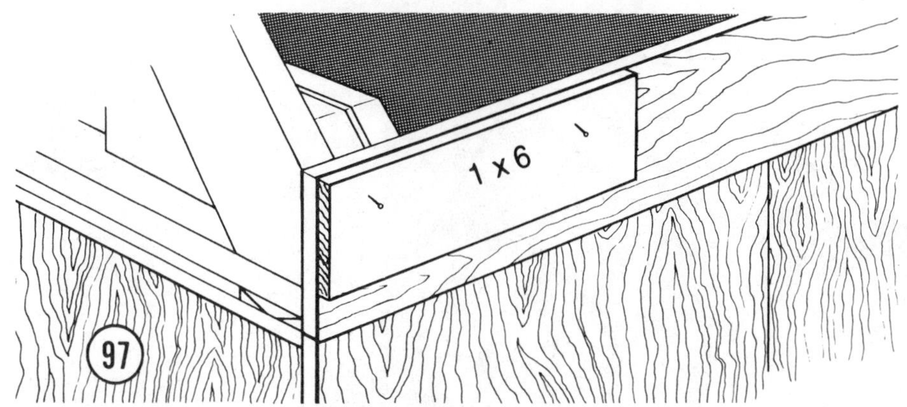

To establish exact amount of overhang, do this. Tack a small piece of 1 x 4 in position, Illus. 97. Consider the 1 x 6 the fascia. Cut a 1 x 2 x 4'. Allow end of 1 x 2 to project 1" beyond top edge of fascia. Consider the 1 x 2 x 4' a 4 x 8 panel of roof sheathing. With 1 x 2 projecting 1" over face of fascia, mark end on gable rafter. Measure same distance up from end of other gable rafter and snap a chalk line, Illus. 98. Nail first course of roof sheathing along this line.

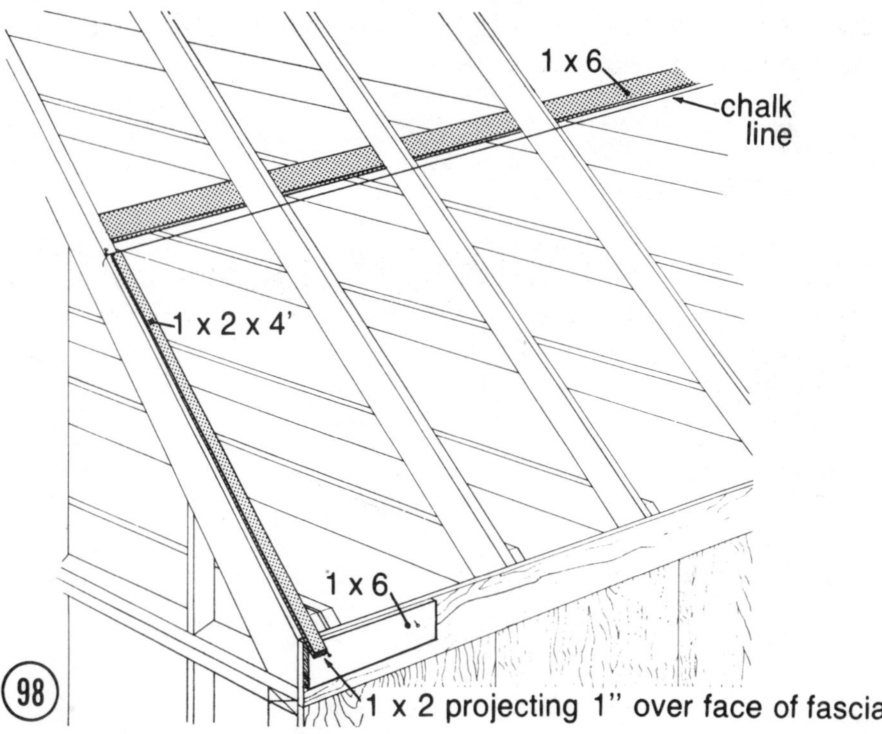

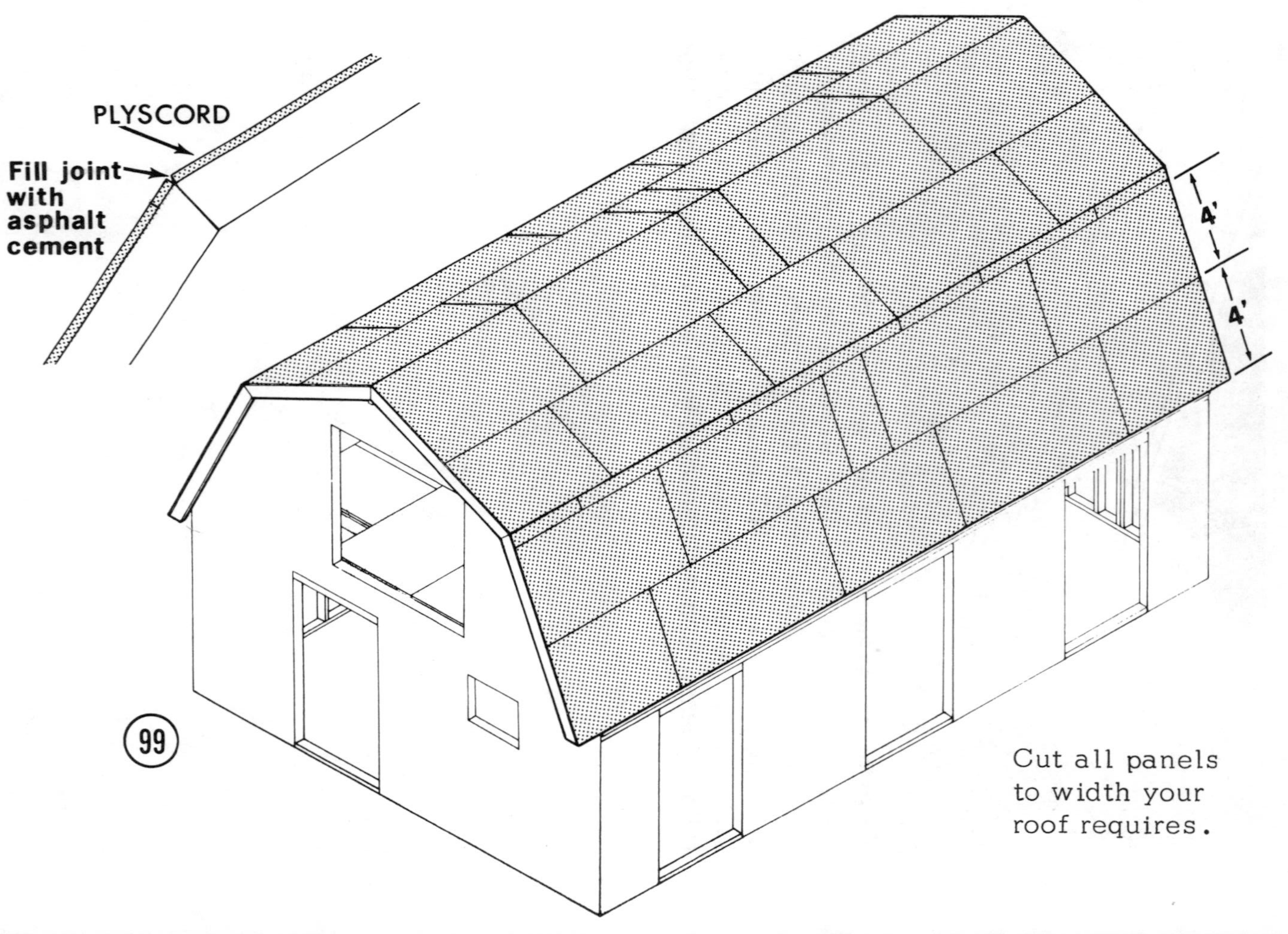
PLYSCORD
Fill joint
with
asphalt
cement
4'
4'
99
Cut all panels
to width your
roof requires.

Stagger joints in roof sheathing. Illus. 99, provides a schedule for applying roof sheathing. When you reach the 1 x 6 temporary braces, remove same. When you reach the ridge, saw or plane edge to shape shown, Illus. 100.

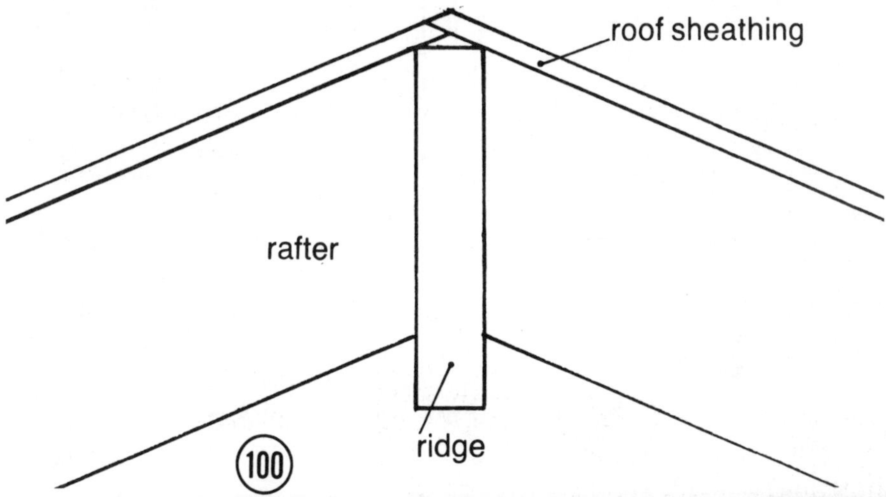

Apply #15 felt to roof if metal roofing is applied. See page 103 for additional information. Always follow roofing manufacturer's directions.

You can now spike 2 x 4 x 8 collar beams D, Illus. 76, 80, in position. To simplify nailing, cut two 2 x 4 E to length needed. Tack these plumb to rafter with one nail. Use same 2 x 4 posts to place all D at same height.

FASCIA BOARD

4—1 x 6 x 12
4—1 x 6 x 16

Use clear 1 x 6 for fascia board on gable ends, and on sides. Cut ridge end of gable fascia board to same angle and length as rafter B, plus ¾", Illus. 74. Cut 1 x 6 fascia to same length as an A rafter, Illus. 73. Nail a wood gusset plate to inside face to reinforce joint. With one person holding bottom end, nail fascia to center end of ridge, Illus. 101, then nail through plyscord sheathing into fascia. Do the same on other half of roof.

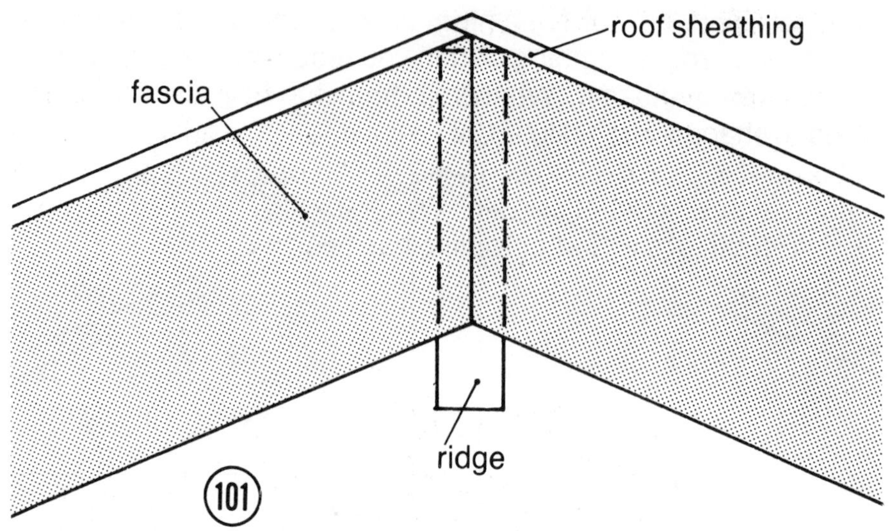

Cut and nail 1 x 6 clear lumber for side fascia. Side fascia butts against roof sheathing, Illus. 102. Nail fascia in position indicated.

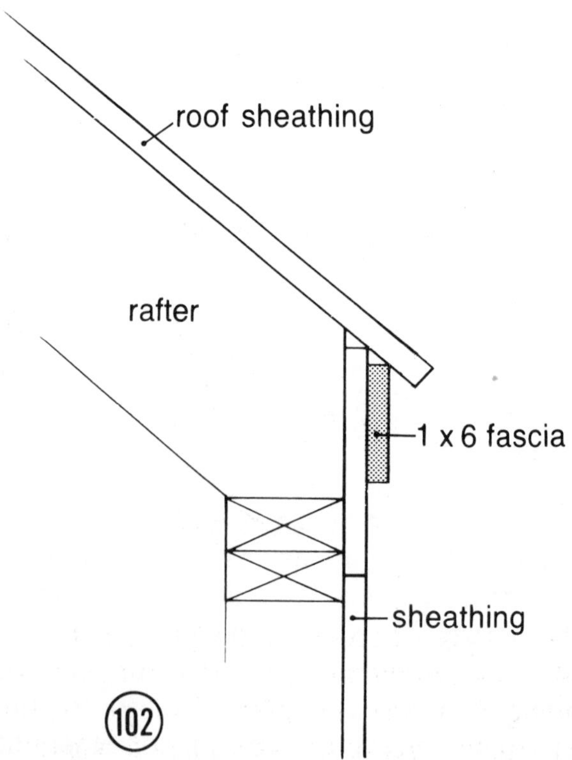

GUTTER AND LEADERS

There are various kinds of gutter brackets available, Illus. 103. Some are attached to fascia, others are nailed to roof after applying felt. Select and install gutter brackets according to manufacturer's directions. Install gutter at pitch retailer suggests to full length of sides after roofing has been applied.

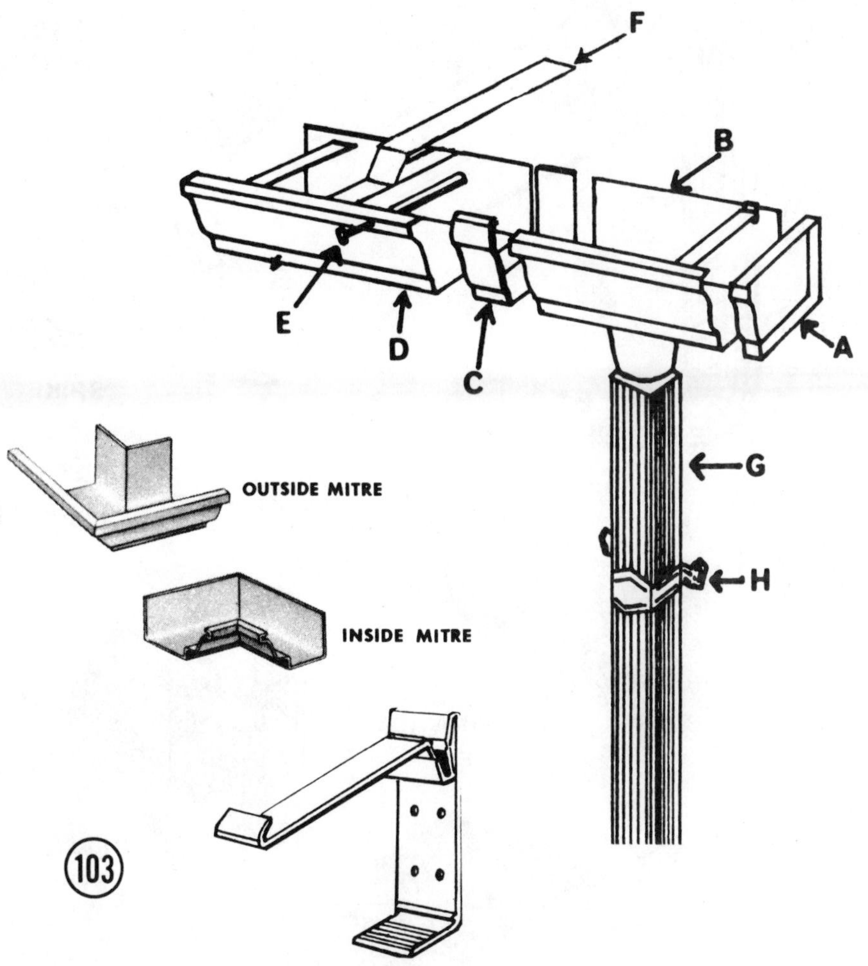

Gutter brackets depending on type used, are fastened from 32" to 64" apart. Additional support is provided aluminum gutters by inserting 7" aluminum spikes E, Illus. 103, and spacer ferrules. The ferrules prevent spikes from buckling gutter.

While Combination Hangers, and/or Fascia Strips, permit installing gutters to fascia after roofing is applied, those who want to install a Roof Apron Strip should nail it to the roof before applying roofing.

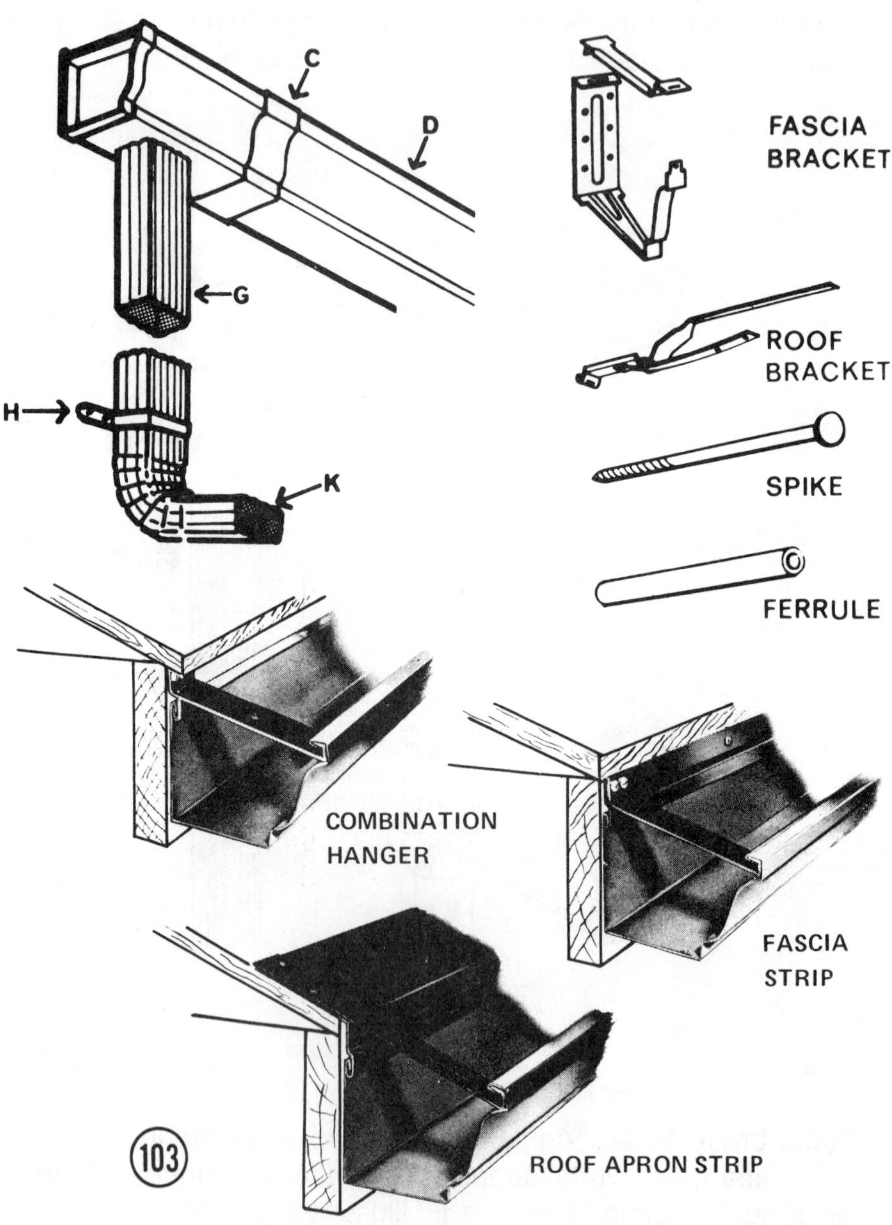

(103)

ROOFING

There are many quality roofing materials available that provide years of carefree coverage. Color coated galvanized cold rolled steel is one. Double coverage mineral surfaced 110 to 120 lb. roll roofing is another.

Double coverage roll roofing is 36" wide. 17" is mineral surfaced, 19" unsurfaced. Since each manufacturer of roll roofing specifies how to apply, it's important to follow their directions.

The first step is to nail a 4" wide strip of copper or aluminum, Illus. 104, to the eave and rake (gable). This acts as a drip cap. Bend edge of metal over to cover edge of sheathing. Nail every 8" to 10" about 1" in from inside edge. Use copper nails with copper, aluminum nails with aluminum.

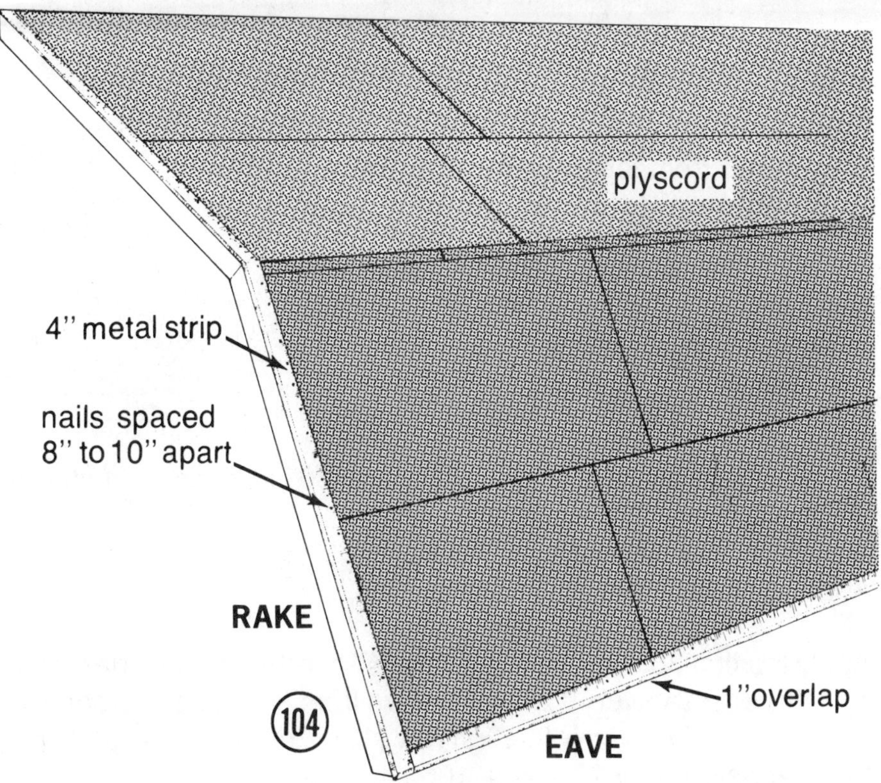

Next cut the 19" unsurfaced portion of roll roofing to length required to cover eave and rake, Illus. 105. Apply asphalt cement and embed this 19" strip to roof allowing it to project ⅜" over eave and rake. Nail to roof every 12", about 4½" down from top edge, 1" up from bottom edge, plus another course of nails about halfway between. Use ⅞ or 1" No. 11 or 12 gauge big head roofing nails. Some roofing manufacturers recommend using hot asphalt, others cold asphalt adhesive.

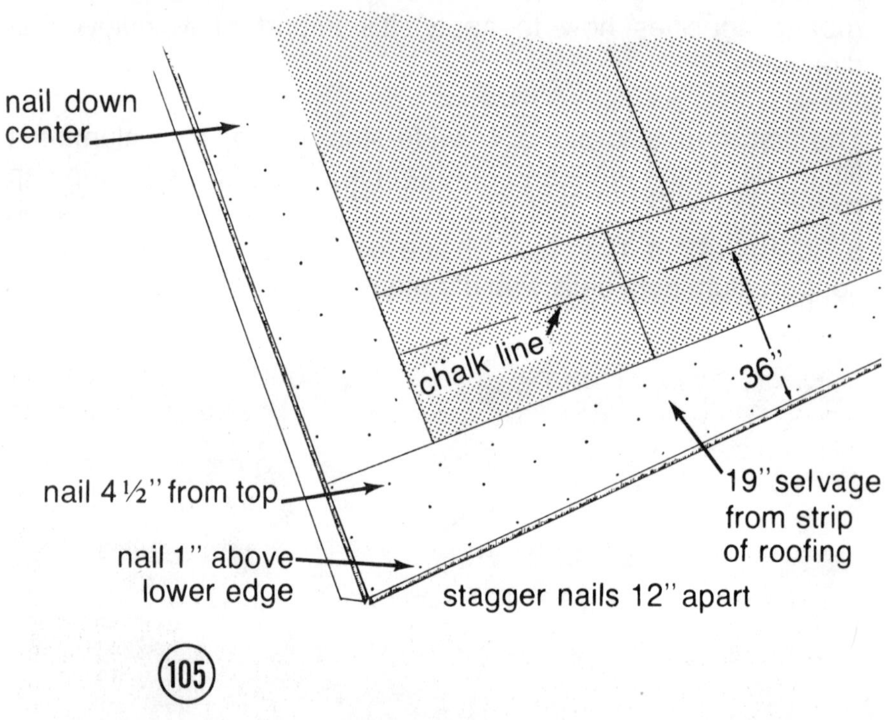

To apply first course, snap a chalk line 36" from edge of starter strip.

Apply asphalt cement over starter strip.

Apply the first 36" course flush with ⅜" overhang on starter strip. Nail this course every 12", 4¾" down from top edge, the second row of nails 8½" below the first row (13¼" down from top edge). Stagger nails every 12", Illus. 106.

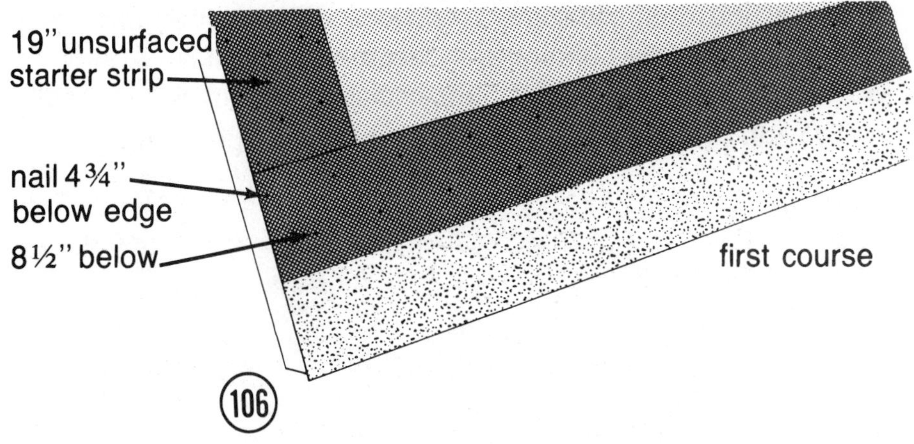

Apply cement to unsurfaced area of first course. Lay second course of roofing in position shown, Illus. 106B. Repeat nailing procedure outlined previously.

When laying the second and succeeding courses, don't butt mineral surfaced edge of one course up to mineral surfaced area on previous course. Allow ¼" separation. This permits rolling second course onto first course without having the mineral surface louse up the bond.

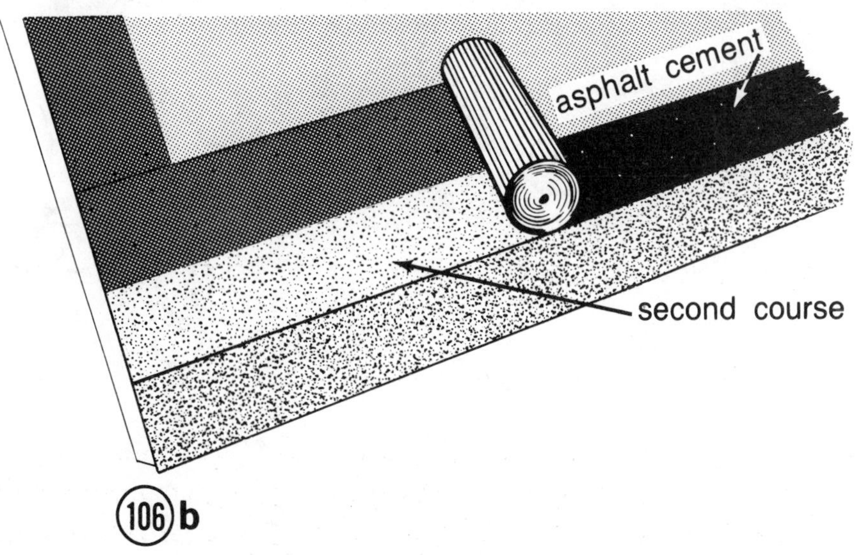

After applying roofing to ridge, finish ridge following manufacturer's directions. Some manufacturer's recommend cutting a strip of roofing 12" wide to length required. Cover 5" of ridge with asphalt, embed and nail ridge strip ¾" up from bottom edge, every 2".

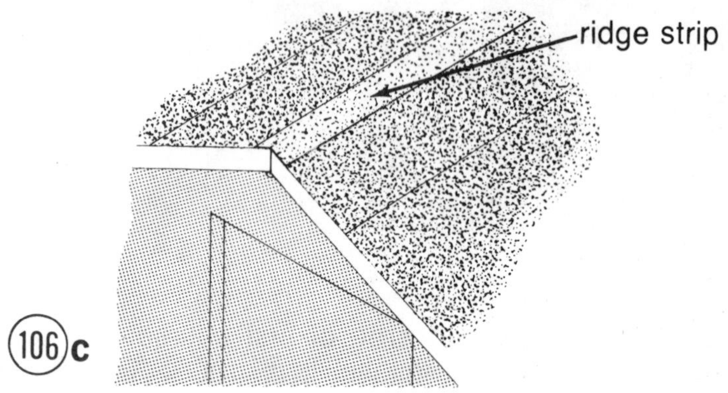

While anyone who isn't afraid of height can apply roofing following the simplified directions provided by manufacturers, there are certain basic problems that must be resolved before beginning. Book #696 Roofing Simplified explains how to make a body harness, how it's used, as well as working from a scaffold. It also recommends using non-skid, ankle high, rubber soled Keds.

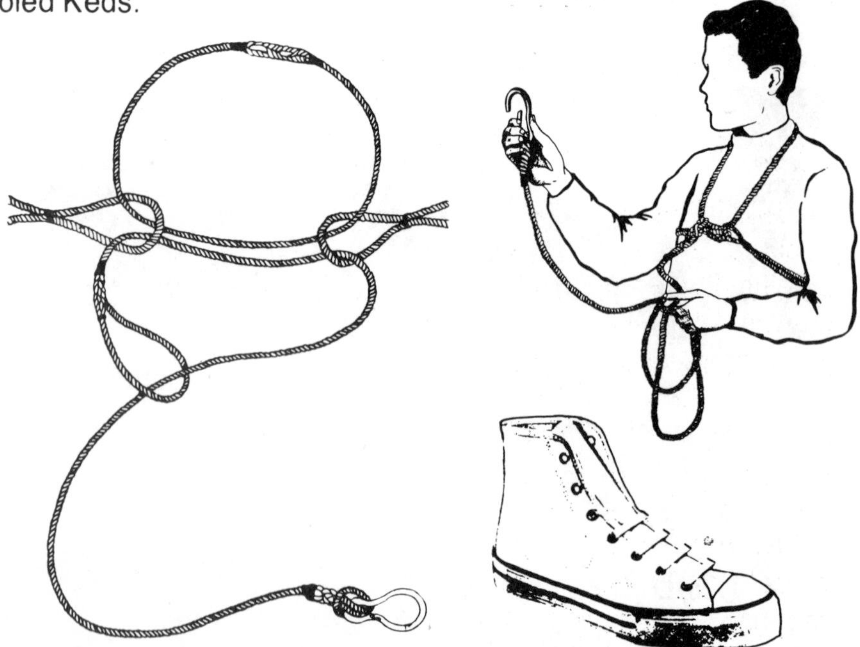

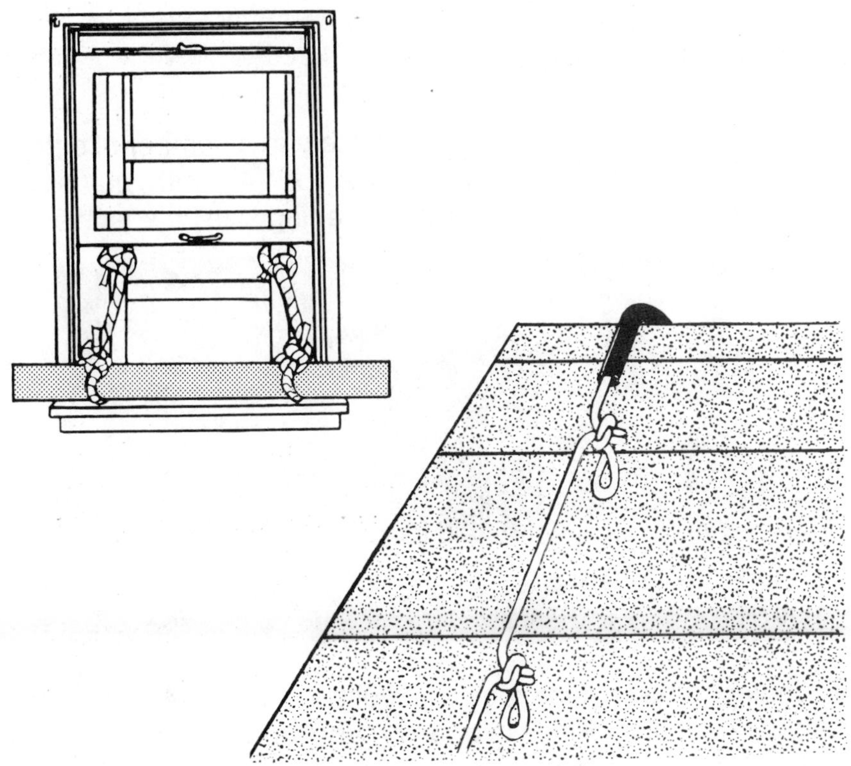

An adjustable pipe scaffold, Illus. 64, should be raised to height needed to nail metal drip cap to rake and eave.

While an extension ladder can be laid on the roof and held in place with a ½" or larger nylon line laid over the ridge and secured to a 2 x 4 placed across a door or window, before placing ladder on roof, lash two 18" pieces of 2 x 4 across bottom edge. This raises ladder off roof, permits unrolling 36" wide roofing and working over it.

Apply the metal drip strip, the 19" starter course, the first and second courses working from a scaffold built to eave height. Lay the third and other courses working from ladders lashed to roof. When you reach the ridge, bend course over and hold edge in place with 1 x 2, then nail every 4", about 1" up from edge. Start roofing on other side at eave. Remove1 x 2 and lap this course over ridge. Nail in place. Next apply ridge strip, Illus. 106C.

WINDOW FRAMES

5—2'4 x 1'4" windows.

Stock cellar sash, Illus. 107, is available in a number of different sizes. If your retailer doesn't stock 2'4 x 1'4" sash, buy size he sells and frame openings to rough opening size he specifies.

107

To make frames to size required, cut two 1 x 6 for A, Illus. 108, to height of opening. Place in opening and measure length for 1 x 6 B. Cut B to length required. Cut 2 x 8 for C. Overall length of C equals B, plus 1½" (two A), plus width of two 1 x 4 casings D, Illus. 110.

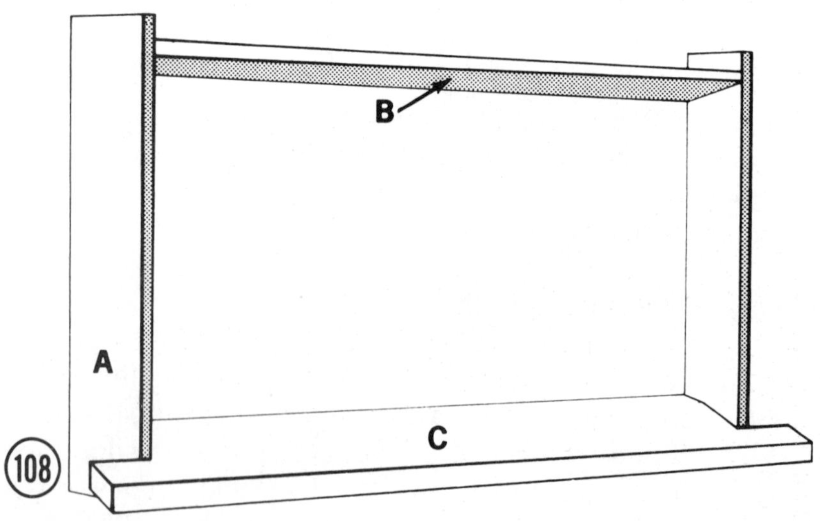

108

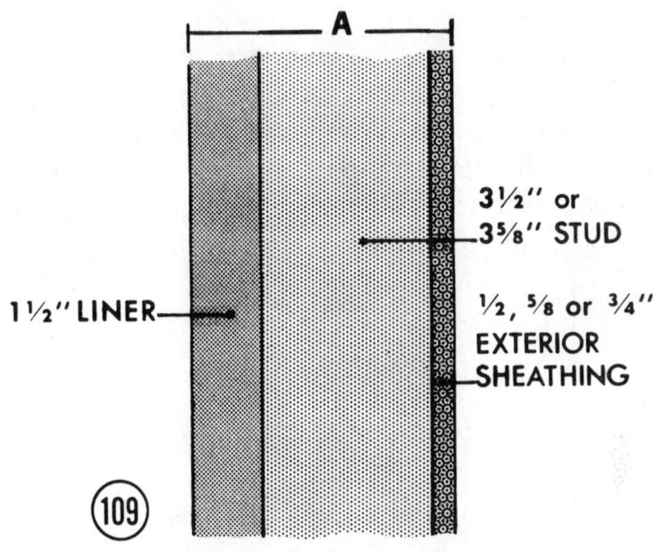

A and B, Illus. 109, is always as wide as total thickness of exterior sheathing (⅝"); stud (3½"); and inside liner (1½"). Recess casings ⅜" from edge of A and B to give that professional look, Illus. 110.

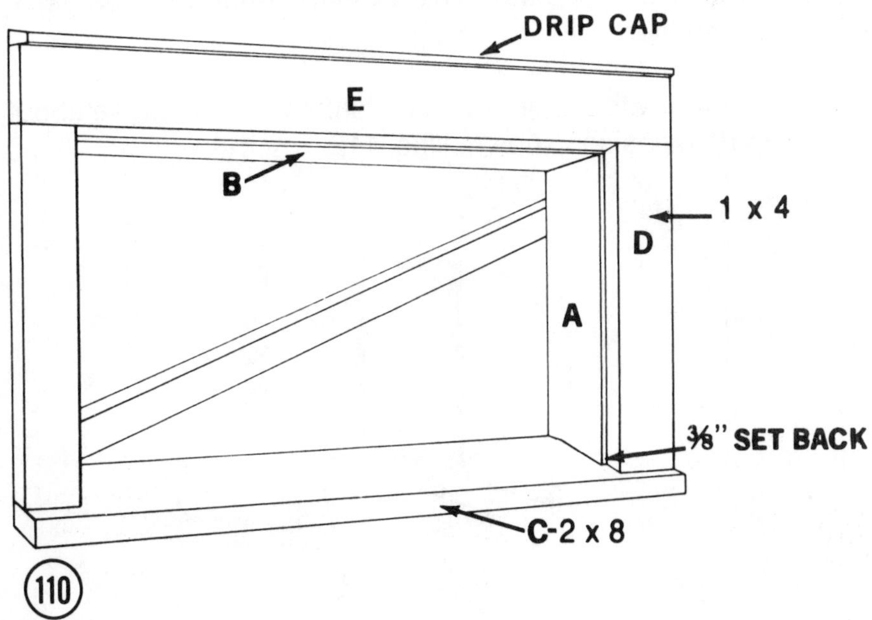

Cut sill C to shape shown, Illus. 111, and to length your window requires. Plane front edge of C perpendicular. Plane base for window level. Apply glue and nail A to B and C. C projects beyond opening and finishes flush with 1 x 4 casing, Illus. 110. Check frame with square. When square, nail diagonal 1 x 2 bracing across AB. Nail casings D and E, in position. Nail drip cap. Paint frame with two coats of outside paint before installing in position.

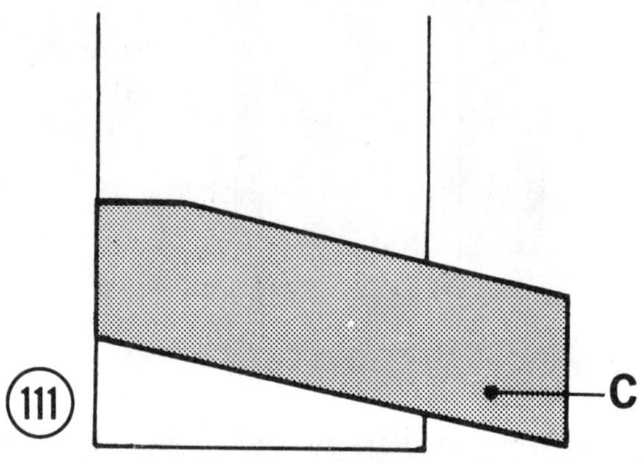

Place frame in opening. Check with level. When plumb and level, nail through C into header with16-penny common nails. Countersink heads and fill holes with putty. Frame projects 1½" beyond inside edge of studs. This permits frame to finish flush with 2 x 8 stall liner.

Install windows with a pair of 3" butt hinges. Fasten these about 3" in from B, in position shown, Illus. 112.

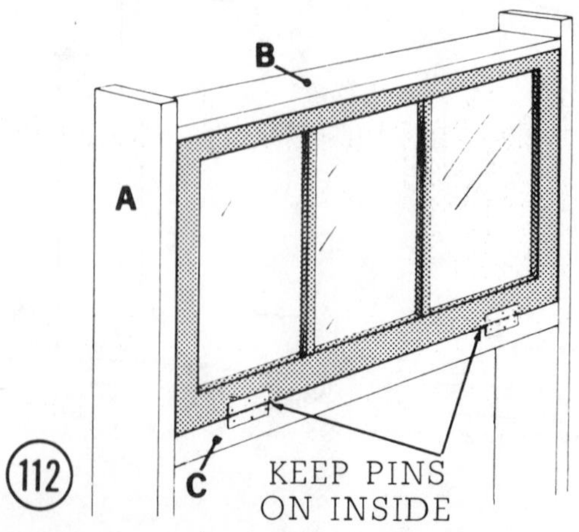

DOOR FRAMES—OUTSIDE STALL DOORS

Cut shoe off across door opening. Use 2 x 6 to build a door frame, Illus. 113. Cut sides A, to full length of opening. Glue and nail A to B. Check and hold square with 1 x 2 diagonal brace. Nail 1 x 4 casings C and D in position shown, Illus. 114. Paint frame with two coats of outside paint or apply wood preservative.

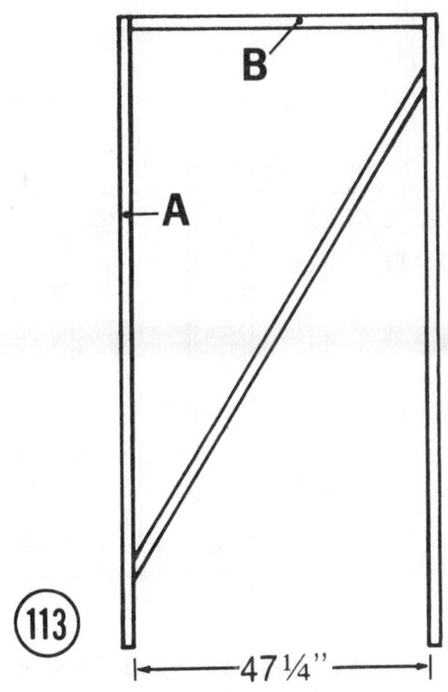

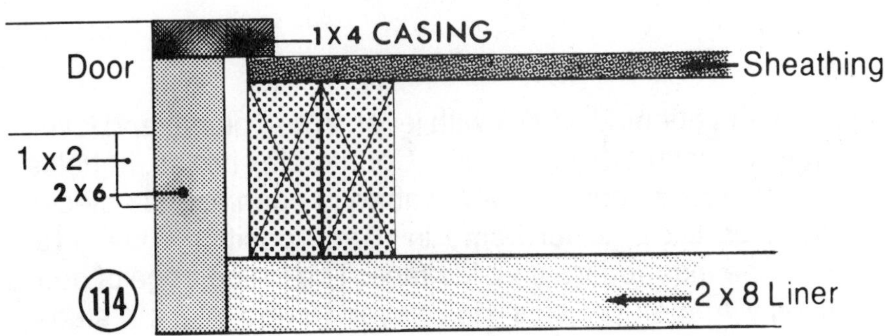

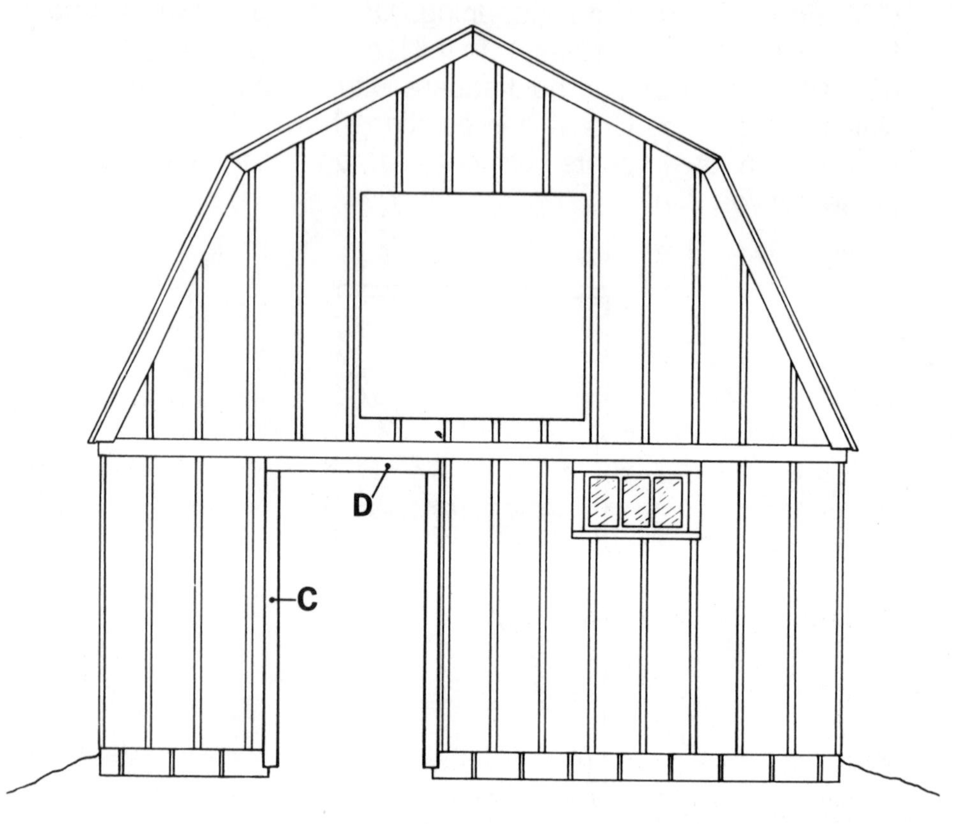

(114) To add colonial charm, nail 3/4 x 2-1/8 battens , 16" on centers.

Place frame in opening, check with level. When plumb and level, nail in position with 16 penny nails. Countersink heads, fill holes with putty. If frame needs to be plumbed in opening, use pieces of shingle, or 1 x 2, to fur frame in plumb position, Illus. 115. These can be inserted from inside face. The inside edge of door and window frames projects 1½" beyond edge of 2 x 4 studs. This permits 2 x 8 or 2 x 10 stall liner that actually measures 1½" thick to butt against frame.

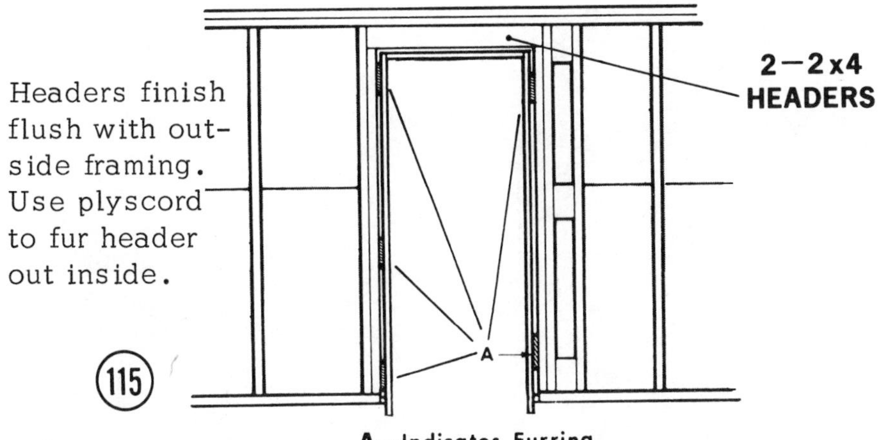

Headers finish flush with outside framing. Use plyscord to fur header out inside.
2—2x4 HEADERS
A
115

A—Indicates Furring

Outside stall doors should be dutch doors, Illus. 16. Check overall width of opening and build a door ¼" less in overall width and height. Use clear 2 x 6. Since these are heavy doors and subject the frames to considerable strain, it's necessary to nail the frames securely in plumb position.

BUILD OUTSIDE STALL DOOR

A professionally constructed one piece or dutch stall door, Illus. 16, can be made in two ways. The first requires gluing up 2 x 4 and six 2 x 8's for an inner, kick-proof lining, Illus. 116, then applying a 1 x 6, or 5/4 x 6 frame on the outside, Illus. 117.

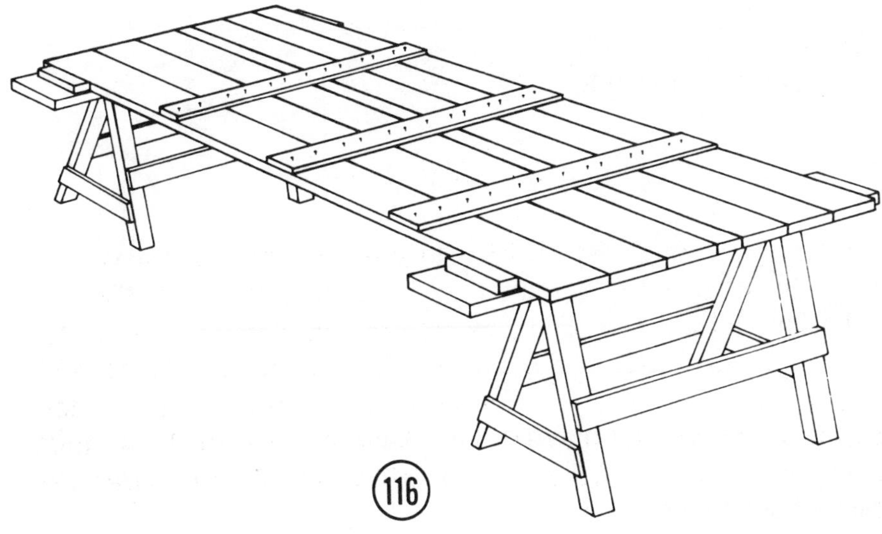

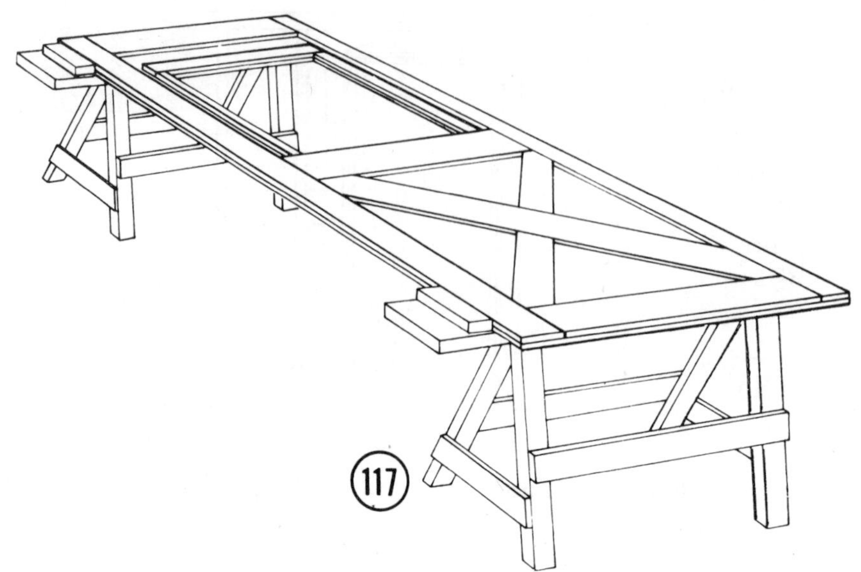

A quicker method suggests using ¾ x 4 x 8 exterior grade plywood and 1 x 6 framing as shown in Illus. 118. The one piece door can be hinged or installed with sliding door track.

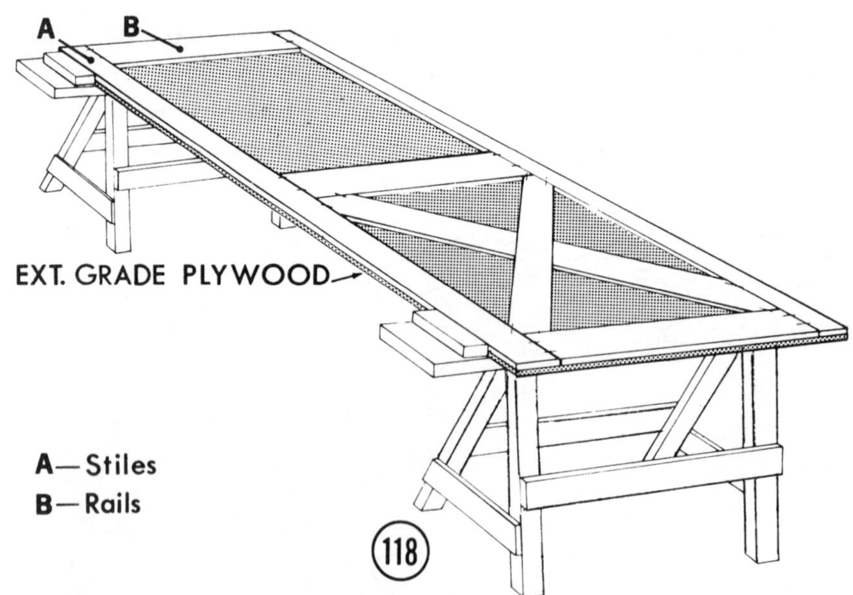

To build a kick-proof dutch or solid door, nail two 2 x 6 x 6' to the top of two sawhorses, Illus. 119. Nail two 2 x 4 blocks to the 2 x 6 in position shown.

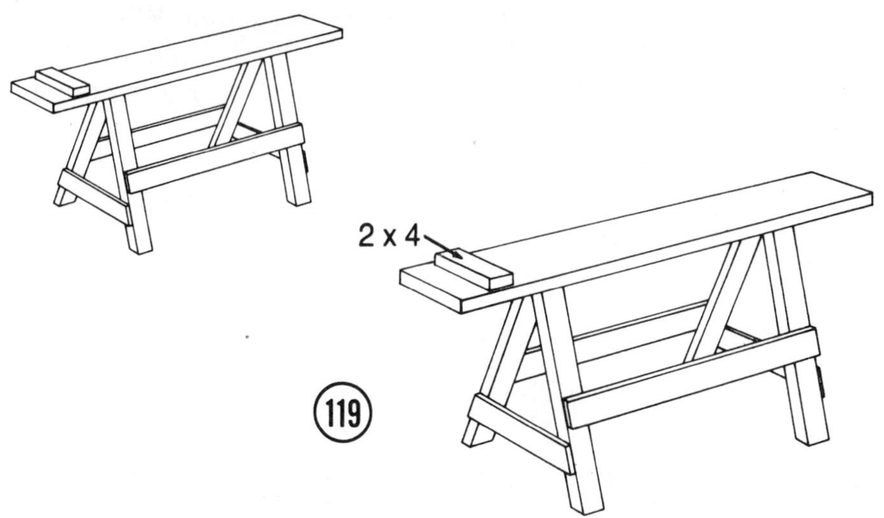

Cut 2 x 8 x 8', plus one 2 x 4 x 8' to length required. Since 2 x 8 presently measures 7¼", you will need six, plus one 2 x 4. This presently measures 3½". This adds up to a 47" wide door. The 47¼" opening in door frame allows ⅛" clearance on both sides.

Apply glue to edges of 2 x 8 and clamp first two planks together, Illus. 120.

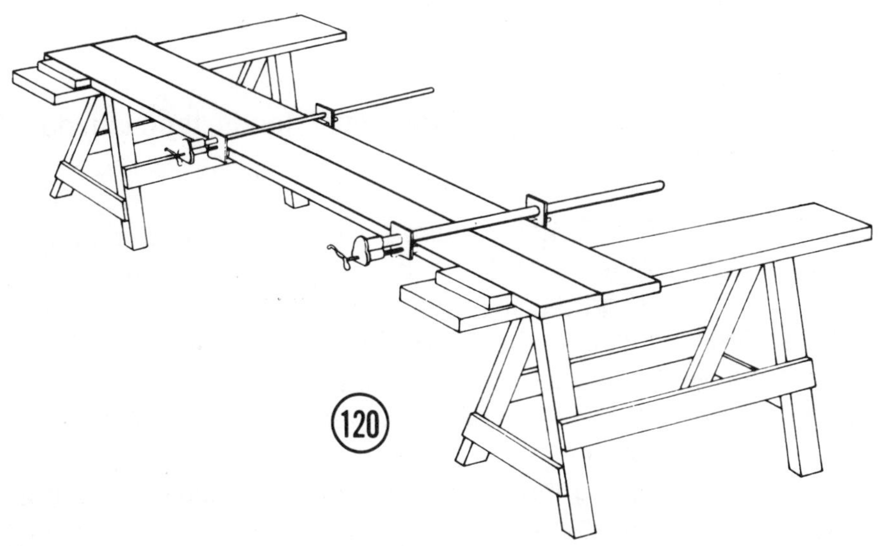

Build door to size your opening requires

Nail 1 x 6's across in position shown, Illus. 121. Apply glue to edge of next 2 x 8, apply clamps to draw it together. When tight, nail 1 x 6. Assemble inner liner following this procedure. Don't drive nails in 1 x 6 all the way. The 1 x 6 is only a temporary brace.

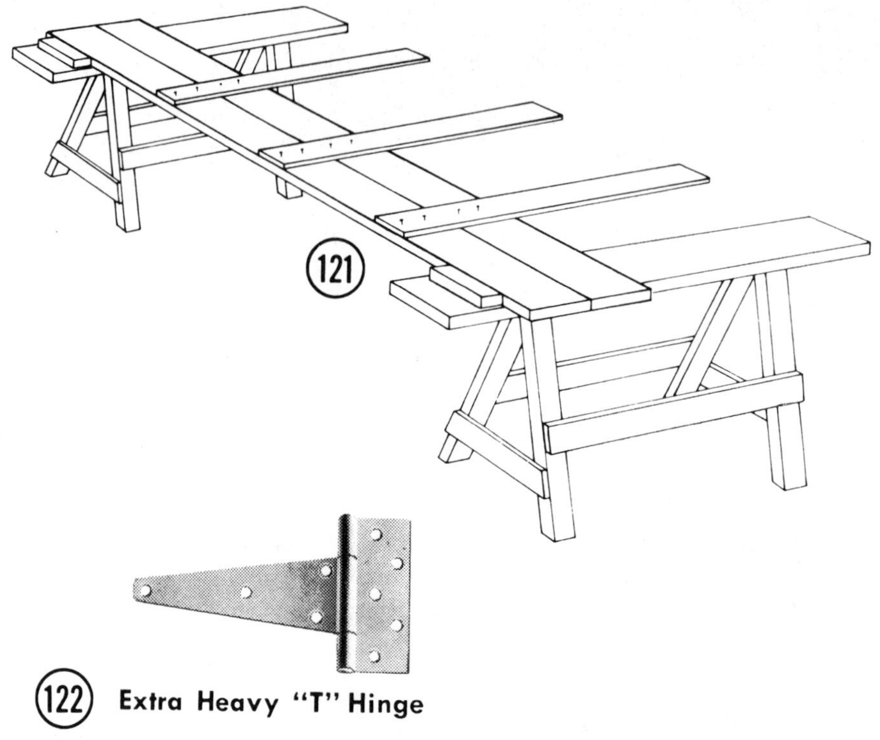

122 **Extra Heavy "T" Hinge**

Allow glue to set time manufacturer recommends, then turn door over and apply front frame, Illus. 117.

Cut 1 x 6 stiles A to length required. Cut 1 x 6 rails B to length required. Apply glue to ends of rails and fasten in position to stiles using corrugated fasteners. Allow glue to set time manufacturer specifies, then turn frame over. Apply glue to entire surface of stiles and rails and fasten frame with 1¾, #10 flathead wood screws. Countersink heads and fill holes with putty. Paint door on outside. Remove 1 x 6 braces and creosote liner. Hang door using three 8" or 10" T hinges, Illus. 122, on a solid door; four hinges on a dutch door. Fasten hinges in center of rails. Nail 1 x 2 stops to frame in position door requires, Illus. 114.

DUTCH STALL DOOR

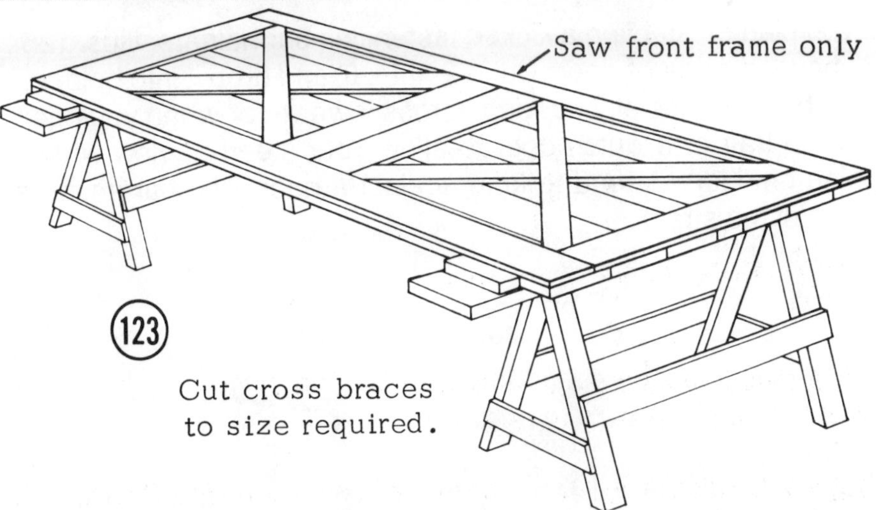

To build a dutch stall door, use 2 x 8 liner as previously described. Build front frame with an extra rail, Illus. 123. Allow glue to set time manufacturer specifies. Saw through front frame only. Saw through liner 1½" above bottom edge of top rail. This permits top half of door to overlap bottom half 1½".

Apply glue and screw front frame to top and bottom liner with 1¾" No. 10 flathead wood screws. Install dutch doors with four 10" T hinges.

Apply door bolts to outside doors, Illus. 124. These bolts work extremely well on stable doors. The one-half inch thick steel handle serves as a pull. It's reversible in its mounting so latch can be used either on a right or left hand door. Cam action holds door securely closed eliminating door rattle. A hole in both ends of handle will accommodate a padlock.

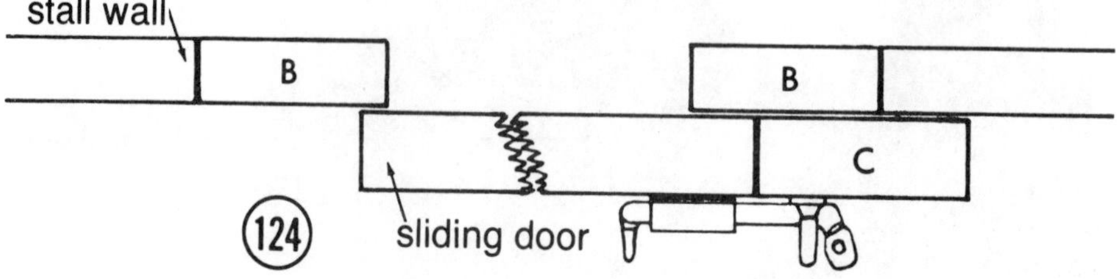

If you live in an area where the horse may be molested at night, apply a second latch to the top half and use a padlock.

If you plan on stabling a kicker, fasten two bar holders, Illus. 125, 6" beyond edge of door. Cut a 2 x 4 x 6' and drop it into holders. This bar is especially needed across lower door when you open the top half of a dutch door to allow your steed to view what's going on. Horses tend to lean against a door. The bar absorbs much of the strain.

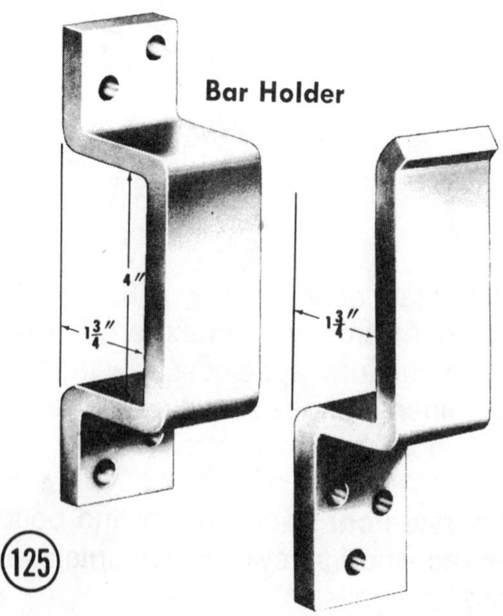

STALL CONSTRUCTION

You can install stall dividers after placing supporting posts B and C under the main beam, Illus. 36; after all floor joists are in position, Illus. 61; or at this time, Illus. 126.

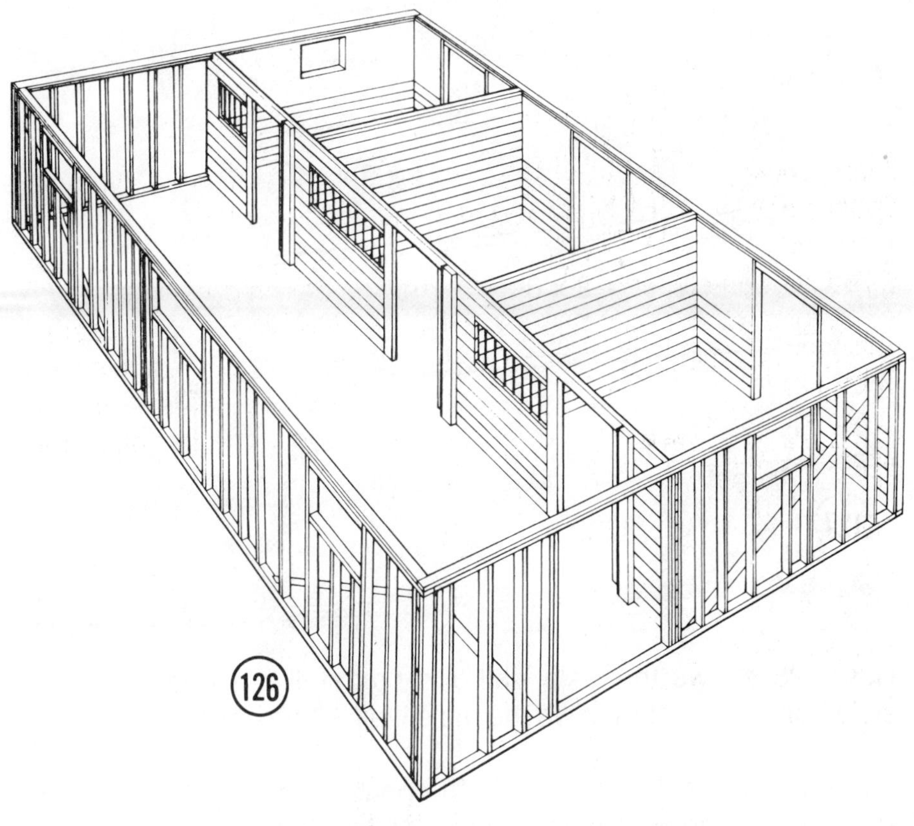

If grilles are to be installed in aisle wall, nail an extra 2 x 8 by length needed under beam A, Illus. 36, 126. Toenail 2 x 8 to A and B. Install 2 x 8 stall boards to a height that allows between 24" to 28" opening for a grille.

Some architectural publications suggest separating 2 x 8 or 2 x 10 planks in stall walls with spacers, Illus. 127. The idea is to provide a freer circulation of air. This type of construction is popular in the south. While it does provide more air, it encourages a horse to develop one of the worst possible habits, cribbing, or wind sucking. By offering so many exposed, chewable edges, it practically encourages a horse to chew wood. We recommend solid stall wall dividers.

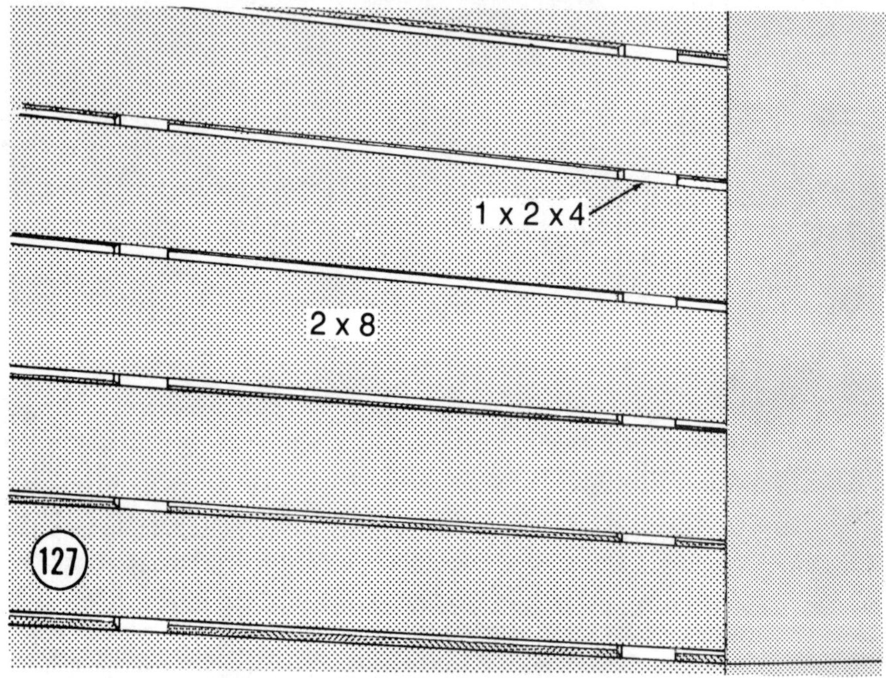

Illus. 128 shows the aisle wall of a box stall boarded up to the aisle beam. Illus. 36 shows the aisle wall with a grille.

There are two schools of thought concerning the installation of stall lining. One suggests nailing 2 x 8 liner to outside walls to within four feet from ceiling joists. A 1 x 2 is then nailed across studs above edge of liner. Another 1 x 2 is nailed to plate. The 1 x 2 furs out edge of plate and studs, Illus. 129. ¾—4 x 8 plyscord panels are then nailed over the 1 x 2 furring. This provides a smooth stall lining. Make cutouts for windows so panels butt against window frames. Nail plyscord liner every 6" along edge of panels, every 8" on studs.

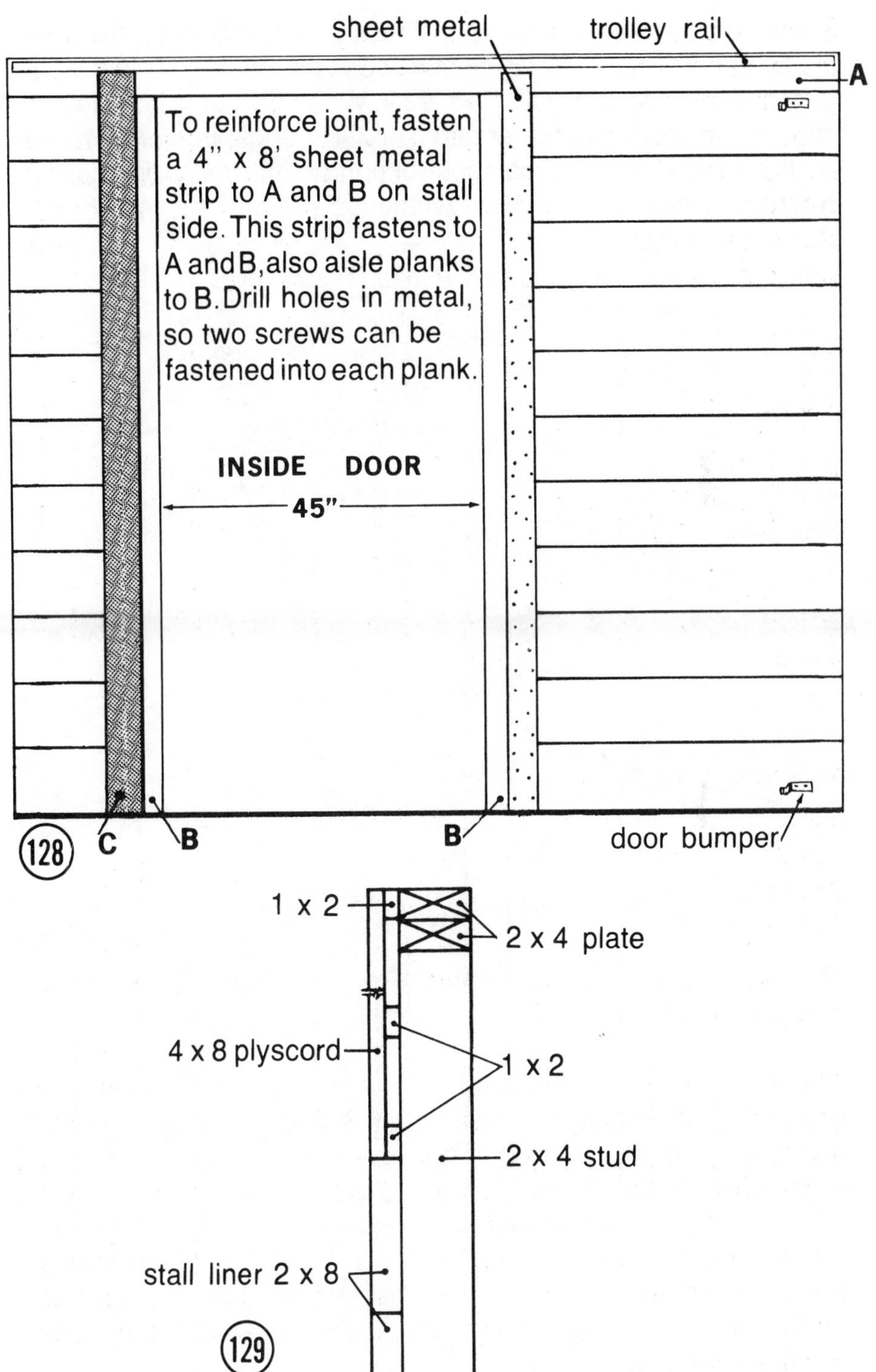
sheet metal
trolley rail
A
To reinforce joint, fasten a 4" x 8' sheet metal strip to A and B on stall side. This strip fastens to A and B, also aisle planks to B. Drill holes in metal, so two screws can be fastened into each plank.
INSIDE DOOR
45"
128
C
B
B
door bumper
1 x 2
2 x 4 plate
4 x 8 plyscord
1 x 2
2 x 4 stud
stall liner 2 x 8
129

1½ x 1½ x 8' angle iron is fastened in position stall divider requires with ¼ x 1¼" lag screws. Secure angle iron to divider boards, with liner on wall, Illus. 130. When divider is boarded up clear to ceiling joists, a second 1½ x 1½ angle iron is fastened on the opposite side. This construction permits replacing divider boards if same should become necessary.

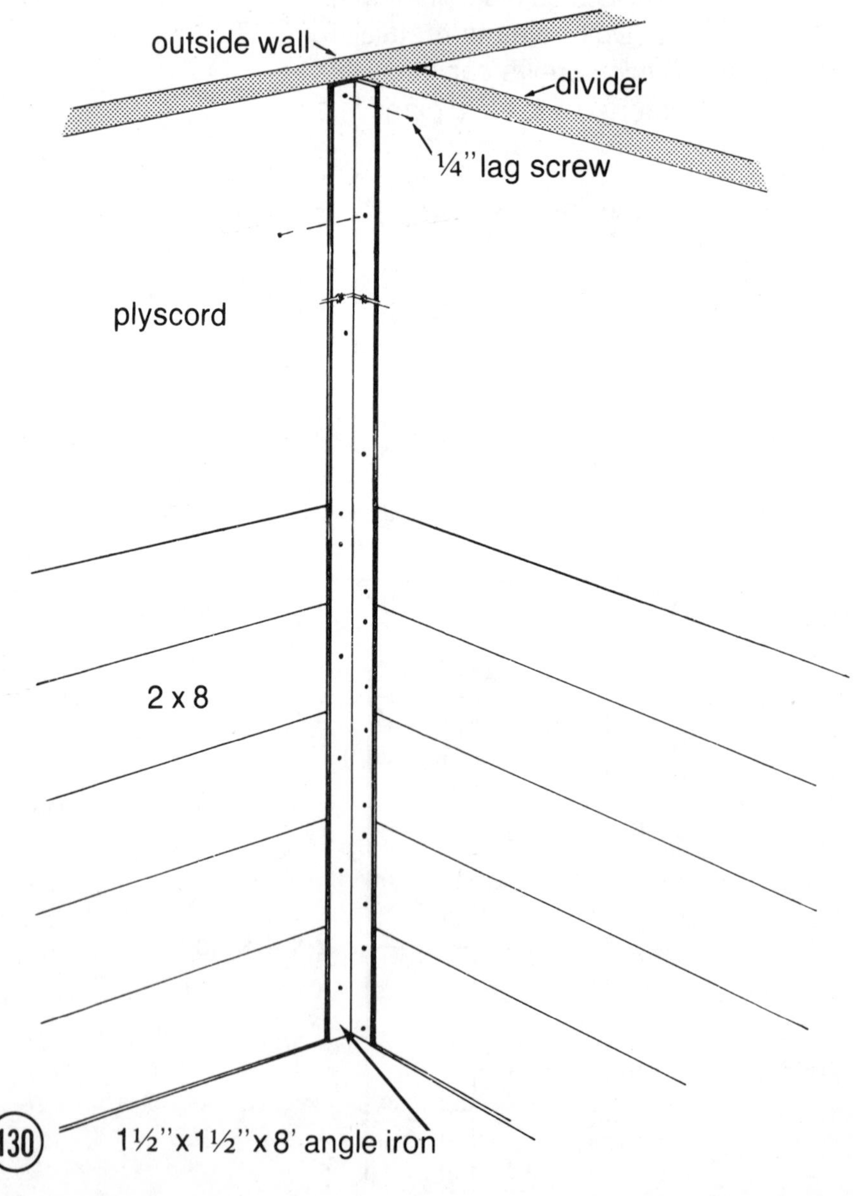

If anchor clips were embedded in footing, nail these to bottom planks with 6 or 8 penny nails. If U-channel was used, notch lower plank to receive head of lag screws. Fasten side of U-channel to bottom plank with ¼ x 1" lag screws.

The second method of installing stall dividers suggests building dividers before applying 2 x 8 lining to outside walls. The divider planks are nailed to a stud in the outside wall. By nailing through aisle planking, ends of divider planks are held in position until 1½ x 1½ angle irons are fastened on both sides. The 2 x 8 stall lining is then nailed to studs. 1 x 2 is nailed to studs above 2 x 8 lining. ¾ plyscord is applied as described previously.

From years of experience the author suggests the first method, applying 2 x 8 stall liner to within 4'0" from ceiling joists and lining the outside walls with plyscord above 2 x 8.

Lumber for stalls, particularly the bottom stall liners, should be all weather pressure treated planks.

Use lumber free of cracks or knots in stall dividers. A solid wall prevents one horse from bothering another, teaching a bad stable trick, spreading a cough or other disease. A solid wall eliminates the need for one horse to fight off neighbors at meal times. Horses, like people, take instant dislikes, or feel great affection. Horses love their homes. They appreciate its safety, warmth in winter, shade in summer. While a solid wall divider, constructed of 2 x 8, 2 x 10, or 2 x 12, is a must between stalls, a grill on top of a 5'0" wall bordering a grooming aisle is optional. It's recommended since it provides more light, more air and an opportunity for the horse to see what's going on. This sometimes presents a problem when you have one animal who doesn't behave when being groomed. Not only will the bad one act worse when he's got an audience, but will also teach the good one things he shouldn't know. In this situation, which can arise at any time, hang a ¼" plywood panel on the outside of the grille.

GRILLE

Wrought iron stall grilles are available in various stock sizes, Illus. 131; or you can make your own to size required.

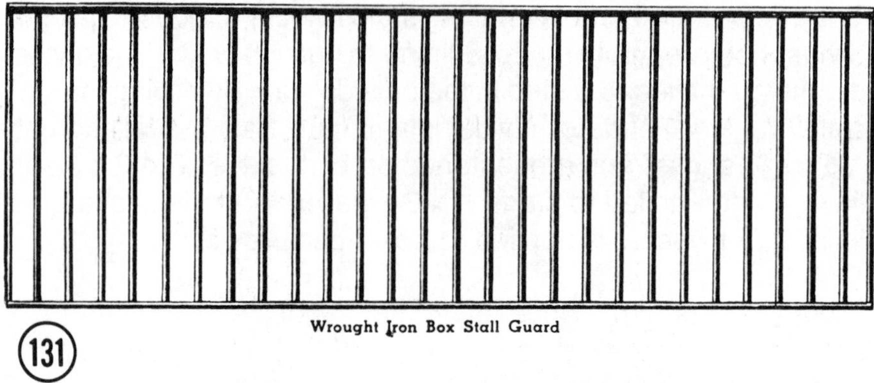

(131) Wrought Iron Box Stall Guard

To make a grille, cut a 2 x 4 to length of opening. Space and drill ½" holes every 3" or 3½", Illus. 132, through the 1½" edge of a 2 x 4. If you have a drill press, it's easy to drill straight 3½" holes (width of 2 x 4). Saw 2 x 4 in half and you have two grille pipe holders with holes identically spaced.

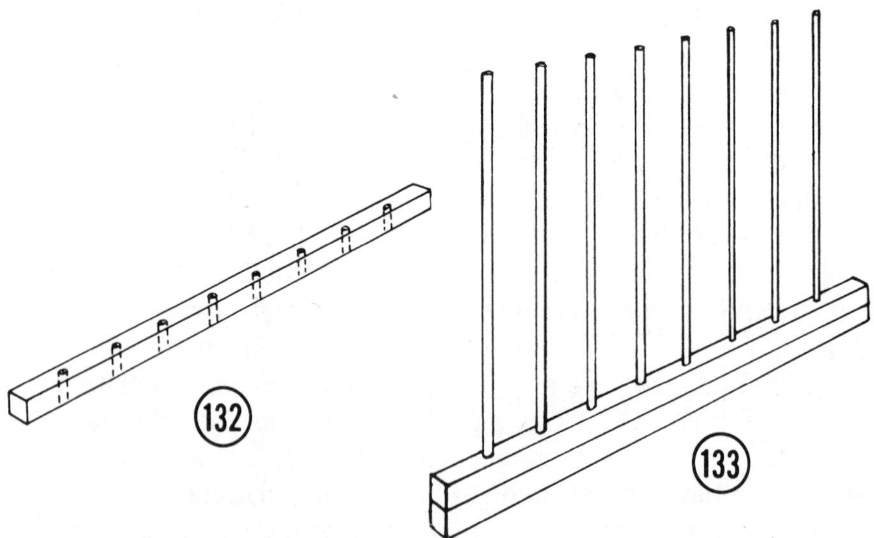

(132) (133)

If you have a hand or electric drill, you might find it easier to only drill through 1½ x 1¾" piece. When you drill one, use it as a pattern to drill the other.

Cut ½" aluminum or galvanized pipe to length required. Tack bottom holder temporarily to edge of 2 x 8. Place pipe through holes in second frame. Place in position over lower frame, Illus. 133. Move top frame into position and screw it in place using 2½" No. 10 flathead wood screws. Predrill screw holes in frame. If a horse kicks and gets cast, bend or hacksaw a pipe and replace it by lowering top frame.

INSTALL HYDRANT, DRAIN, ELECTRIC AND TELEPHONE LINES

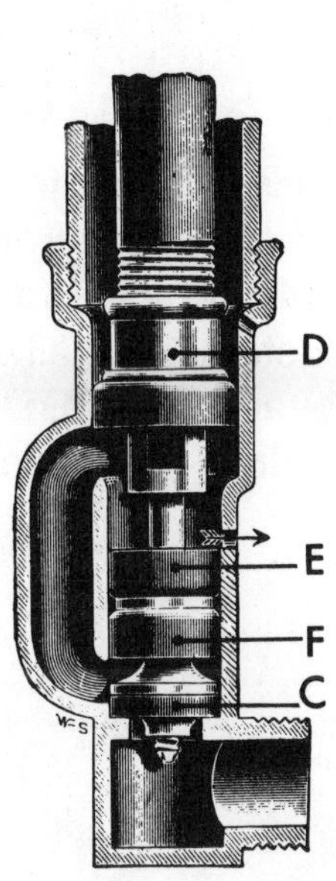

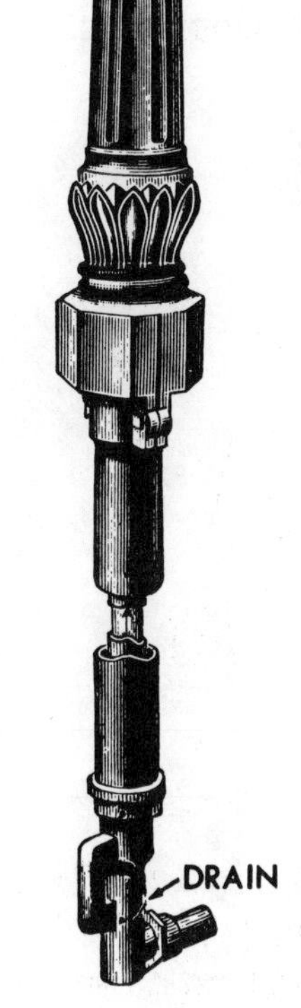

To replace washers, remove spigot A, bolts B. Pull rod straight up. Renew leather washers C, D, E, F.

(134)

Run the water line and connect it to a self-draining hydrant, Illus. 134. Place the hydrant near a post C on the aisle wall. Order a hydrant of sufficient length so its foot valve goes down well below frost level. Non-freezing, self-draining hydrants are available with pipe size connections ranging from ¾", 1" to 1¼". A ¾" connection permits connecting to ¾" or ½" pipe. Hydrants can be ordered in various lengths ranging from 1½' to 6'. This dimension refers to projection below floor level.

Lay 4" drain pipe from a drain directly below spigot. This drain should run to a dry well or run off outside the barn.

PAVE AISLE

Before paving aisle, install and test all water, electric, telephone and drainage lines. Use #8 armored UF cable from house to stable. Install a fuse box alongside the entry door. Run #12 Romex up along ceiling joist to each stable. Install a stall light, Illus. 135 in one corner, preferable over a feed bin. Install only weatherproof outlets in a stable. Run all cable for lights alongside ceiling joists.

135

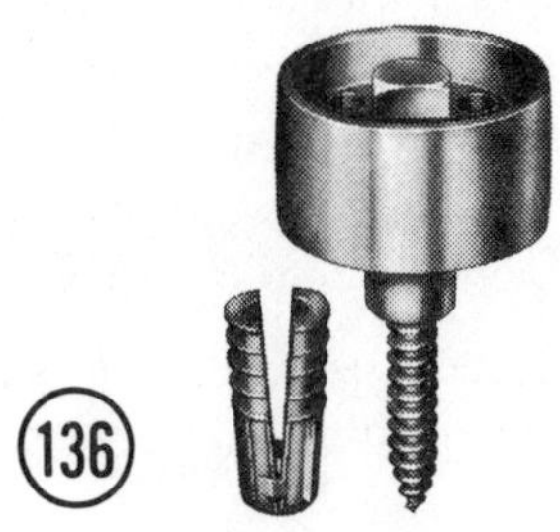

While inside stall sliding doors can be hung after aisle is paved, expansion shields for sliding door floor guides, Illus. 136 should be embedded in position shown, Illus. 137. Floor guides are placed thickness of sliding door plus ¼" away from post B, and in position indicated when door is closed. If you don't wish to embed shield in concrete, a second type of floor guide, Illus. 138 can be screwed to post B before nailing C in position. Drill holes through metal plate so you can fasten these guides to B on right side.

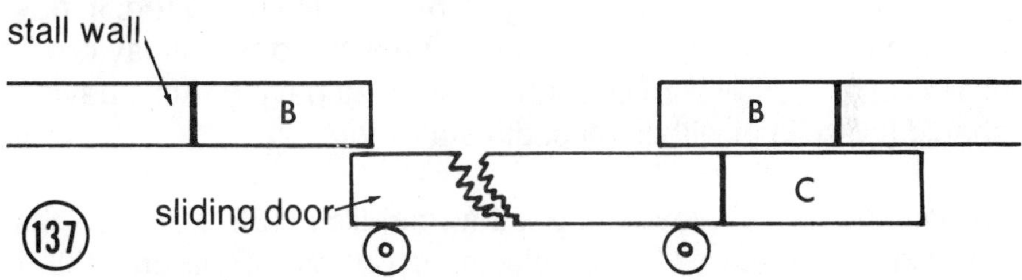

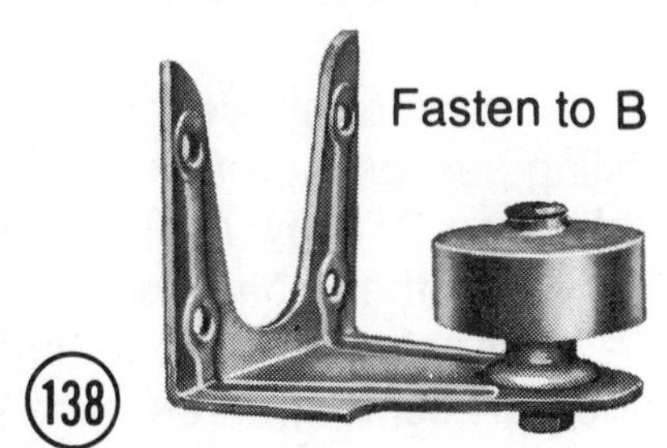

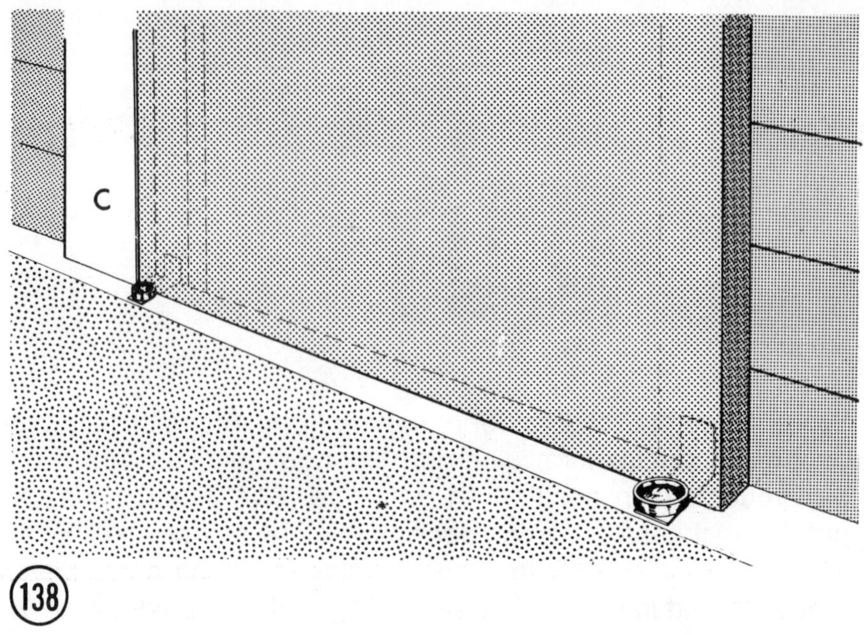

(138)

The underground telephone line should be fastened to the inside of an outside wall and connected to either a wall phone in the aisle, or in the tack room. When water, electric and telephone lines have been brought into position required, it may be necessary to remove some fill in the aisle to permit laying 1" to 1½" of gravel, 1½" of concrete or asphalt paving. Aisle paving should finish flush with level of dirt stall floor.

If concrete is used, level up 2 x 4 forms, Illus. 139. Paint 2 x 4's with old crankcase oil before placing in position. Pour one 4'0" x width of aisle section at a time. Cut 6 x 6 reinforcing wire 3'10" x 84". Spread this over the gravel. Raise it up into the concrete as you lay it. Use globs of concrete to level forms.

Pour section A. Use a 2 x 4 screed to level concrete flush with forms, Illus. 140. After filling section A, remove form A1 and fill void with concrete. Next fill area C. Allow section A and C to set before pouring B and D. When these begin to set, remove B1 and C1 and fill voids. A really smooth surface isn't desirable. Buy ready mix or mix one part cement, three parts sand to five parts ¾" gravel. Your ready mix retailer or building materials retailer sells dust inhibitors. Add this when mixing concrete.

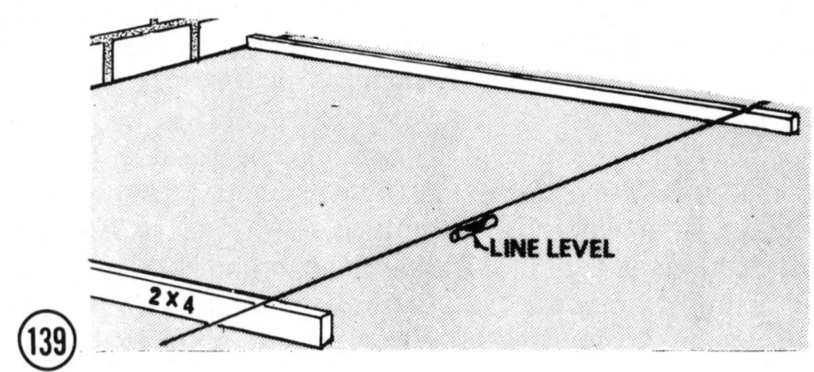

(139)

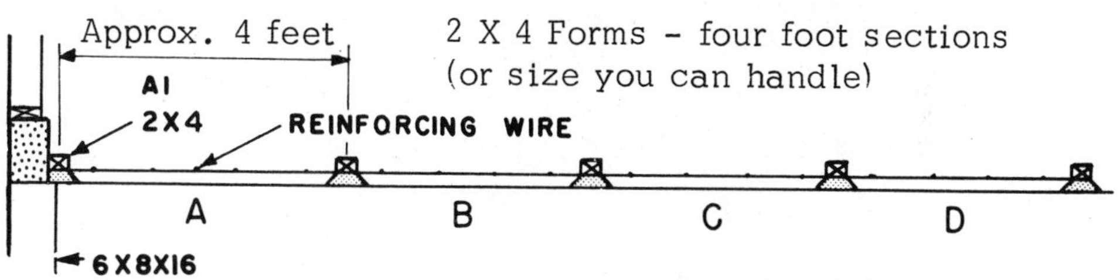

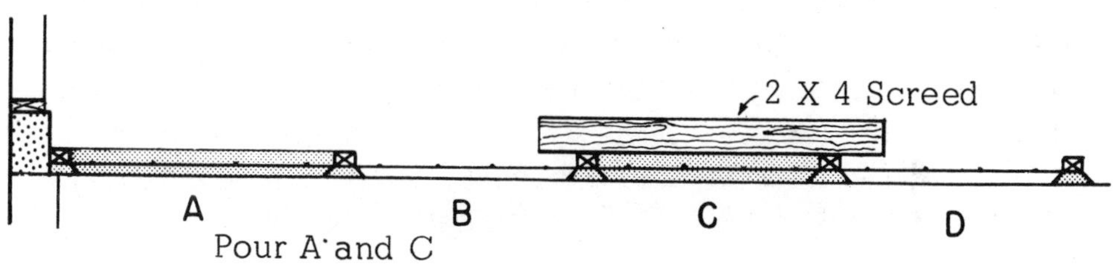

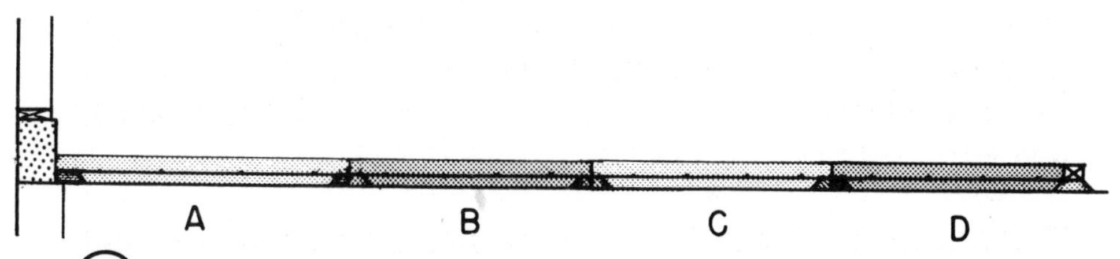

(140) Remove forms and pour sections B and D. Fill in area between section A and curb. Second pouring is indicated with darker shading.

You can embed expansion shields in position required to receive sliding door guide, Illus. 136, or drill holes after concrete sets up using a carbide tipped masonry bit, or use guides that screws into B, Illus. 138.

Many stable owners prefer a dirt floor in the grooming area. While it's less slippery and easier on a horse's hoof, it does require continual dampening to keep the dust down.

If you pave the aisle with asphalt paving, make certain it's compacted with a roller equal in weight to those used on driveways. After allowing a concrete floor to set three days, build a curb around the floor drain under the hydrant, Illus. 141. Drain in curb should be connected to 4" tile running to a dry well or runoff outside of barn. Use 1 part cement to 3 to 4 parts sand.

141

TO BUILD SLIDING STALL DOORS

Use ¾ x 4 x 7' or ¾ x 4 x 8' exterior grade plywood, Illus. 142, plus 1 x 6 front frame. Build door to size of opening plus 2" in width and 1" in height. Saw opening for a stock size grille, or make a grille to fit a 24" high by 28" wide opening. Cut 1½ x 1¾ grille holders to 24" length and install these in door as previously described. A 1 x 2 frame can be nailed around outside of grille. Project 1 x 2 one quarter inch over edge of pipe holder.

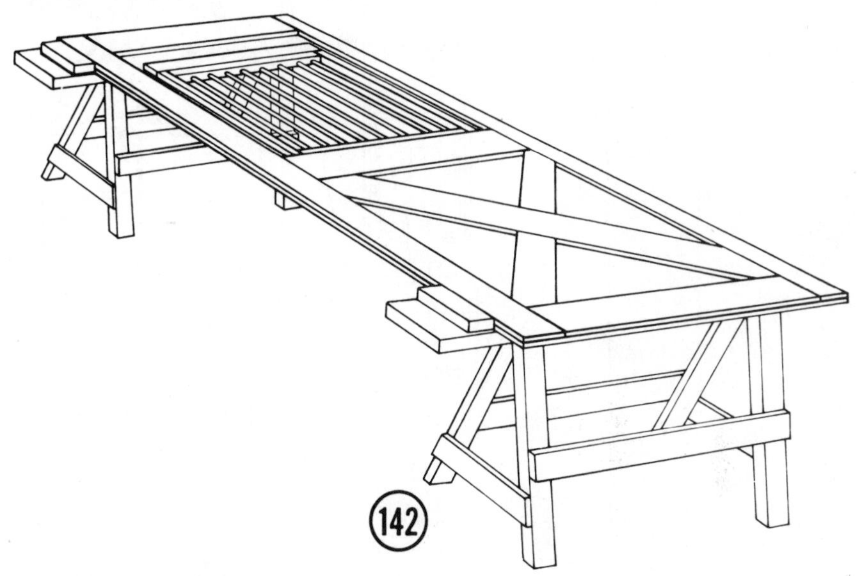

TROLLEY DOOR RAIL

Your building material retailer sells a trolley rail, Illus. 143, designed to handle 1½ or 2" thick sliding doors. Always weigh a completed door (use a bathroom scale) so you can be certain door hardware will accommodate it.

Slide track brackets, Illus. 144, over track. Space and bolt brackets every 24" to aisle beam. Fasten brackets to beam as high up as possible.

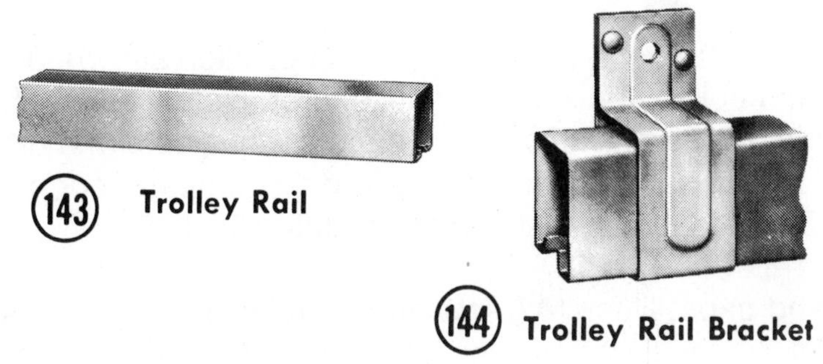

143 Trolley Rail

144 Trolley Rail Bracket

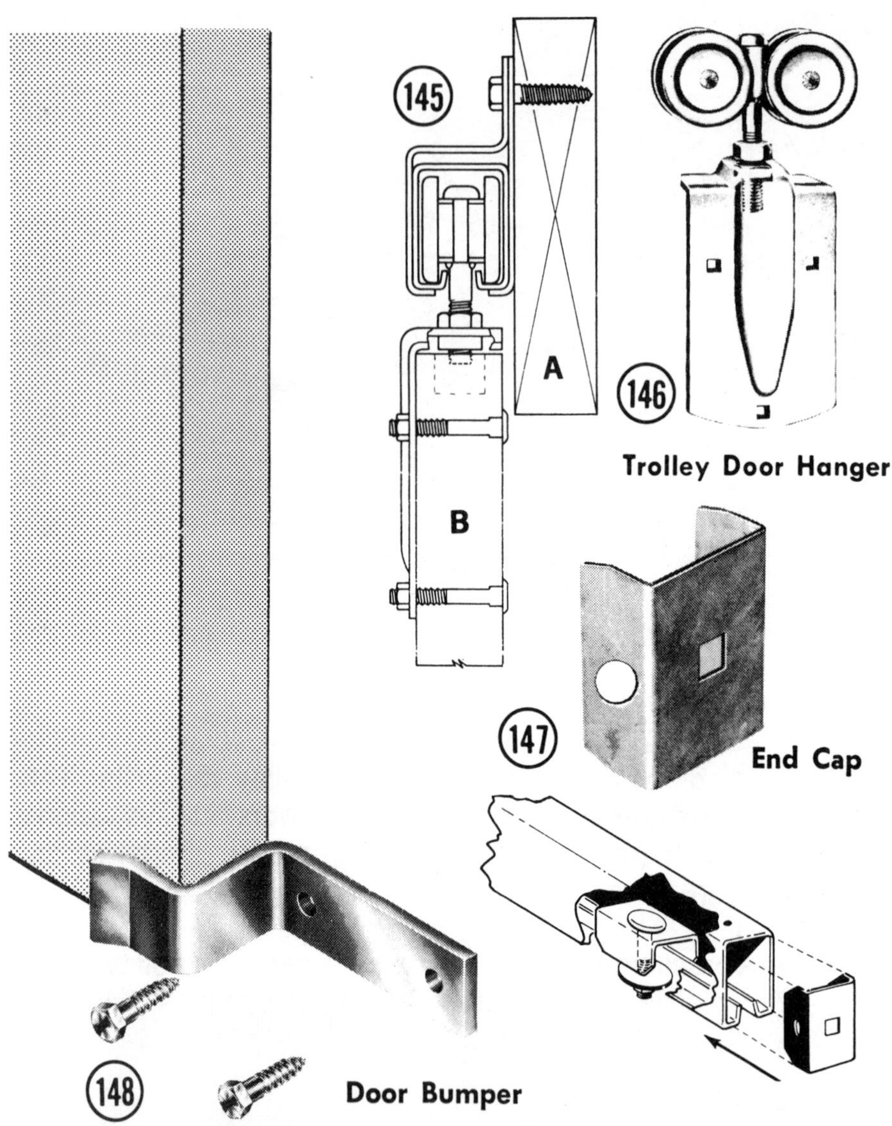

Mount door hangers, Illus. 145, 146, in position shown, 3" from edge, with bolts supplied by manufacturer. Note: Head of bolt is on inside, facing stall. Door hangers are adjustable and provide up to ½" vertical adjustment. If you didn't embed floor guides, Illus. 136, in concrete, fasten guides, Illus. 138, to B.

Drive end caps, Illus. 147, into end of track. These can also be used as center stops, if you follow manufacturer's directions. Fasten two door bumpers, Illus. 148, in position, Illus. 143.

FLOORS

Do not disturb the packed soil in a stall. Always pick up every nail you drop during construction. If codes permit, leave a dirt floor in the passageway until you find you don't like it. If it needs to be paved, use asphalt, rather than concrete. In the days when horses were a vital form of transportation, stable architects specified grooved concrete in all passageways. This was to prevent slipping. In those days, most well heeled owners had stable help that spent time keeping the stable in shape. A grooved floor in a grooming aisle doubles the amount of work. Since a horse is usually cross-tied in the grooming area, and there's no great amount of traffic, asphalt is satisfactory.

Use a wood floor in the tack room, a concrete floor in the feed room.

OAT STORAGE ROOM

Use 2 x 4 for shoe and plate. Space studs 12" or 16" on centers. Use 2 x 4 for ceiling joists. After framing in area, staple ¼" hardware cloth to floor, walls and ceiling. Make 30" door, Illus. 149. Do not cut shoe off at door opening. Cover shoe with hardware cloth. Cover door frame with ¼" hardware cloth. Fasten in place with three 3" T hinges.

149

WINDOW GUARDS

A window guard, Illus. 150, must be installed to keep the horse from breaking the glass. Since stable windows are always hinged at the bottom, guards hold window when open. Use a turn button at top to lock window.

Metal window guards, Illus. 150, are available from stable equipment houses. Build-it-yourself guards, Illus. 151, are easy to build. Cut sides F, Illus. 152, from 1 x 10 or ¾ exterior grade plywood to shape shown and to full height of window frame. Drill holes, screw 1 x 1 aluminum angle to F. Cut ⅛ x 1" aluminum bar stock, or 1 x 2 to length required for bars. Drill holes and screw bar stock to F. If 1 x 2 is used, creosote these and the side guards before installing. Notch F to receive 1 x 2. Screw, do not nail bars to F.

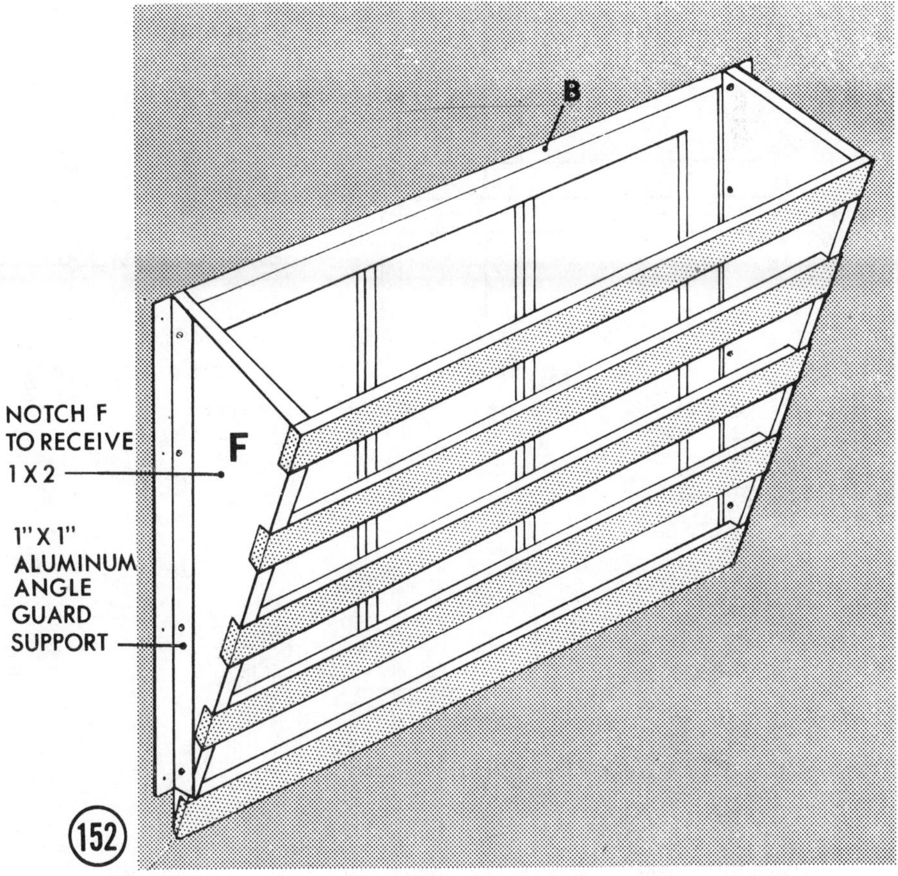

Carefully inspect each stall to make certain no nails or other foreign matter is in any stall. Soak the stall floor several days then allow it to harden thoroughly before laying in a bed of straw.

BUILD HAYLOFT DOORS

Use same exterior grade plywood applied to exterior walls to build doors in loft, Illus. 153. Cut panel to overall size of opening, less ¼" in overall height and width. Saw panel in half to make two doors. Hinge doors with two pairs of 4" T hinges.

LOFT DOOR SILL

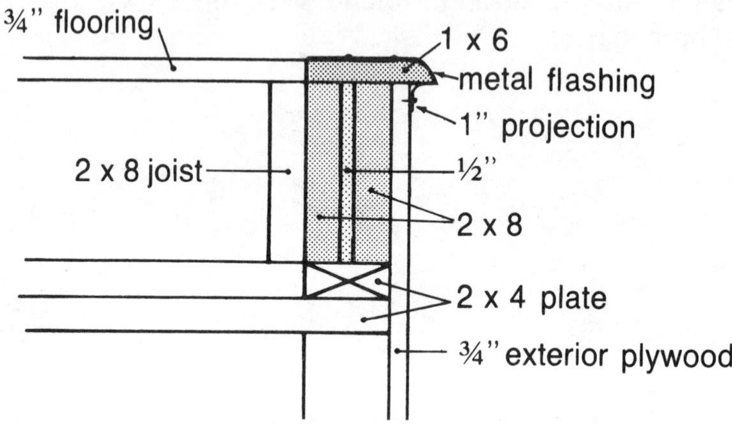

Cut and nail 1 x 6 sill. Cover sill with 8" wide flashing, nail flashing. Bend and nail edge under sill.

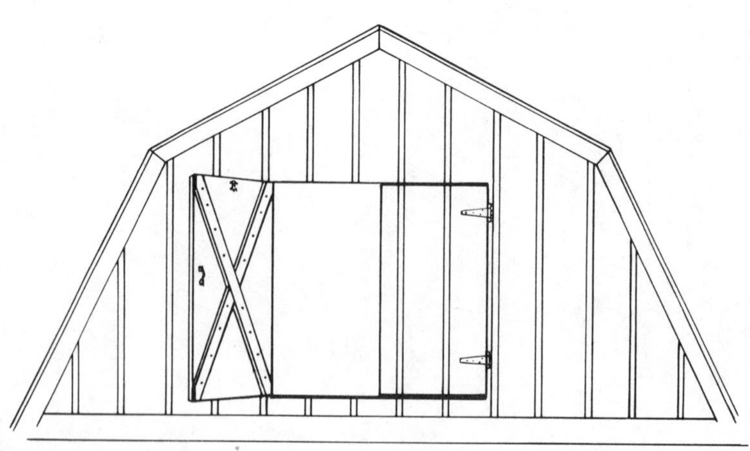

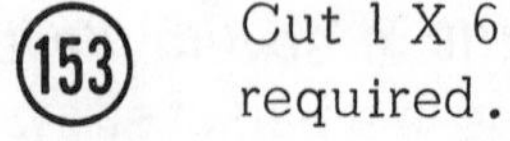

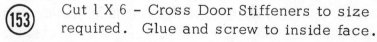

153 Cut 1 X 6 - Cross Door Stiffeners to size required. Glue and screw to inside face.

There is another stable operating procedure that should be considered. If you have young help, one that doesn't mind climbing a ladder once or twice a day, consider whether you want to install overhead hay racks in each stall. These should be placed high up in the corner of each stall and filled through a hole in the loft floor, Illus. 154. Fill the hay racks from 2' x 2' holes in loft floor. To eliminate drafts, hinge doors and keep them closed except when filling rack, Illus. 94.

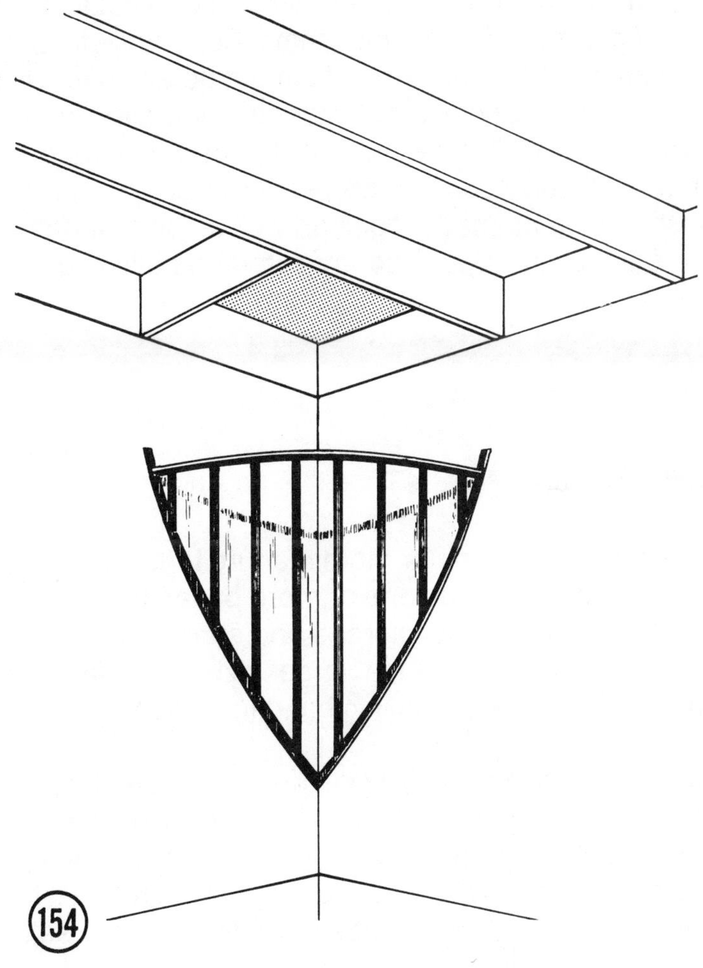

Hay racks are not recommended in colt stalls. A fresh youngster plays in a stall and could get hung up if he kicks his heels through the rack.

APPLY CEILINGS

This is optional. A ceiling does improve the appearance of a barn and helps keep dust in loft from filtering down. A horse needs light and air, but should not be confined in any area that's drafty. Horses are more susceptible to drafts than cold weather. Many owners close a barn up too tight in severe weather hoping to keep the stable warmer. When ceilings are 8'0" or less, there's a good chance of creating condensation, unless a sufficient supply of fresh air is maintained. While fresh air is vital to the health of the animal, drafts must be avoided. Use care to only keep those windows open that provide air without a draft. If you want to install a ceiling, ⅛" tempered hardboard, or ¼" exterior grade plywood can be used. Plyscord can also be used but it should be painted, edges as well as both surfaces, to seal it before installing. Use 6 penny common nails every 12".

STABLE EQUIPMENT

Every stable needs one or more manure forks, a manure basket and a wheelbarrow, plus one or more boxes for grooming tools. It's important to plan tool storage and shelf space into your floor plan. You will also need a large size medicine cabinet for linaments, cough medicine, leg bandages, saddle soap, etc.

Hang manure forks on a wall. Keep manure baskets and a wheelbarrow, or better still, a basket cart, Illus. 155, where it's handy but still out of the way. All parts for these two wheel carts can be purchased KD. You buy plywood locally, cut it to size cart manufacturer specifies, and assemble the cart yourself.

Every stall must have a feed bin, Illus. 156, and water bucket. The feed bin should be fastened securely to the wall 36" from the floor for horses 14 hands and over. Always screw the feed bin securely into a corner.

(155)

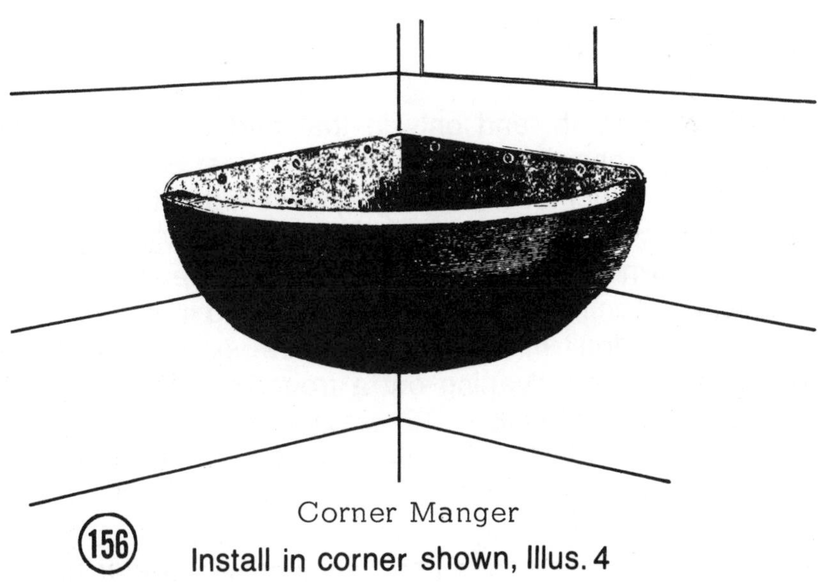

(156)

Corner Manger

Install in corner shown, Illus. 4

Always support a water pail with a 2 x 2 or 2 x 3 crossbrace, or 1" metal strap, in position shown, Illus. 157. Nail 2 x 4 blocks noted. This permits removing and cleaning pail before refilling. A water pail should be cleaned at least once every day. Don't be conned into installing "automatic water troughs" with buried or exposed copper or plastic tubing.

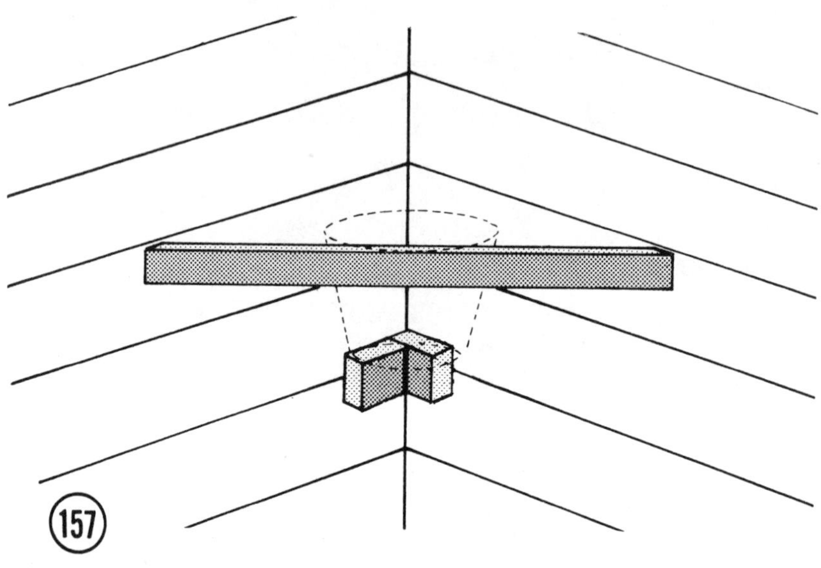

Except in the south, and only in that part of the south that doesn't have freezing weather, like Key West South, can you safely use an automatic watering system without a freeze-up. The automatic watering trough is a great idea. Most architects and builders who don't own horses specify same when drawing up plans, but don't get suckered into installing them unless you like plumbing, don't mind multiple freeze-ups on a cold, cold morning, and enjoy cleaning out a trough with a mean horse breathing down your neck.

While great in concept, i.e., the horse will always have fresh water, in practice the trough seldom gets cleaned. It continually picks up dirt. When a sudden cold spell comes before the lines and troughs have been inspected, a freeze-up creates a big mess. Use pails and place your water in that corner that's most convenient to the sliding door.

If your stable help is inexperienced, or just chicken, you may want to install a feeding door, Illus. 158. Cut out a 12" piece of liner, large enough to put through a 2 quart measure, Illus. 159. Hinge the piece you cut out with two 1½" butterfly hinges. Apply a 3" bolt.

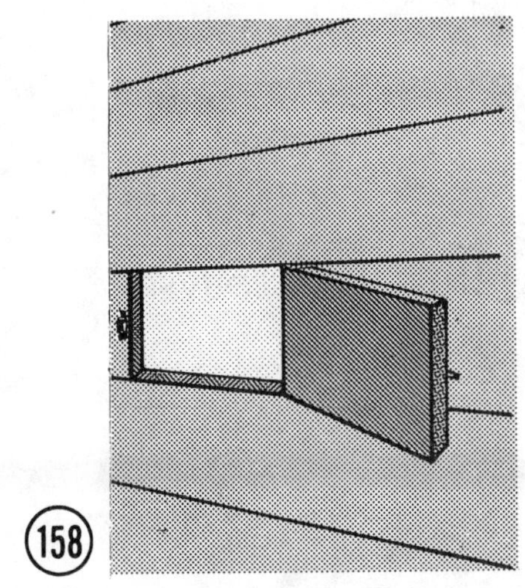

158

159

Install adjustable aluminum louvers in gable ends, Illus. 160. These can be purchased in several sizes. Or roof louvers, Illus. 161, can be installed. Follow manufacturer's directions.

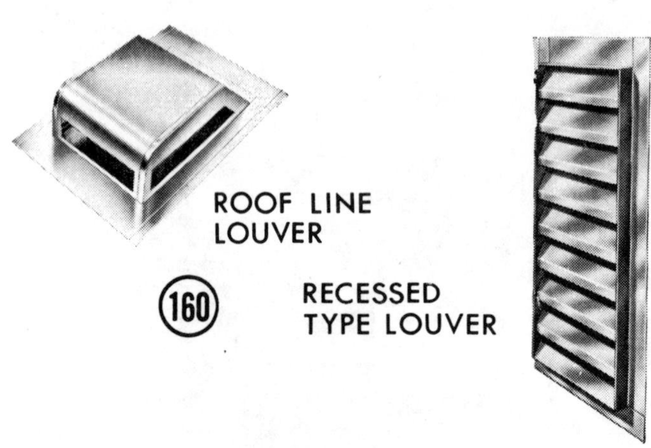

GRAIN BOX

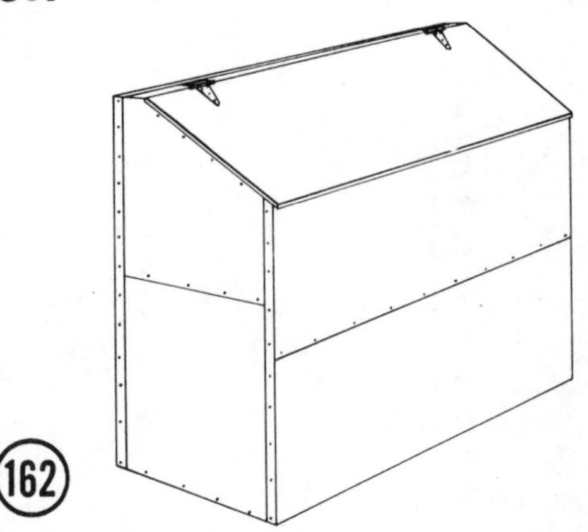

Illus. 162 shows a grain box that measures 36" high in front, 42" at back. Cover with tin or aluminum to provide rat-proof storage for oats. Use ¾ plyscord for all parts. Cut bottom A—18 x 42; ends B—18 x 42 x 36". Glue and nail B to A. Cut back C—43½ x 42. Apply glue and nail C to AB. Nail 2 x 4 in position shown, Illus. 163. Cut front D—43½ x 36". Glue and nail in position. Cut lid size required, project 2" over front. Bevel edge against 2 x 4.

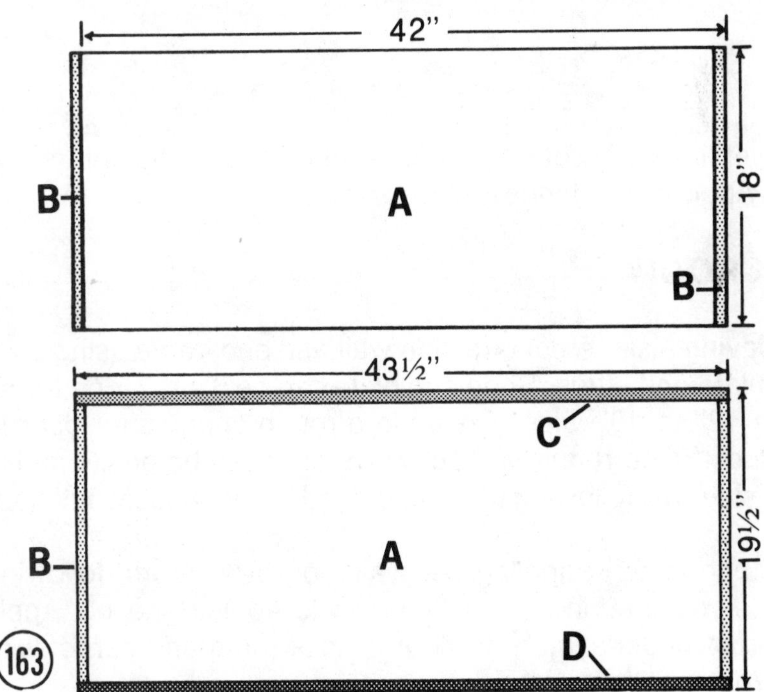

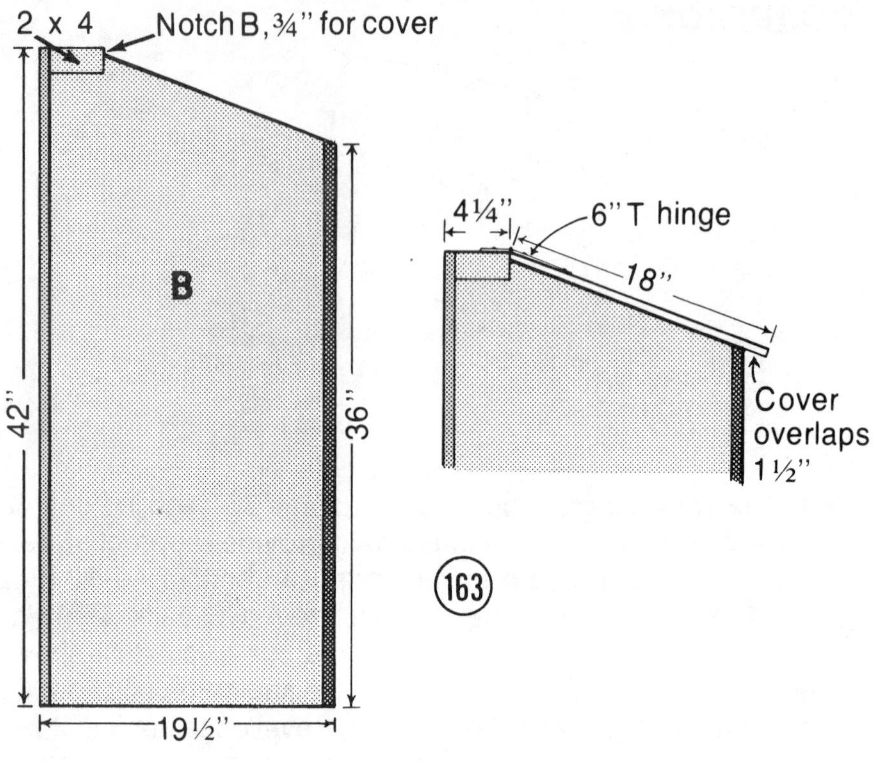

Cover entire box, bottom, sides, ends, lid and top with sheet metal. Apply two 6" hinges to 2 x 4 and lid.

TACK ROOM

After paving aisle, erect one side wall and end frame using 2 x 4 shoe, plate and studs. Shoe and plate can be 6' or 7'. End frame can be 42" to 48" wide. Frame in a rough opening for door to size stock frame requires. Tack room door can be positioned in end or side wall following framing procedure previously outlined.

Apply hardwood paneling to walls of tack room following directions outlined in Book #605 How to Apply Paneling. Apply waterproof underlayment to floor of tack room and carpet with indoor/outdoor carpeting.

TO BUILD A CUPOLA

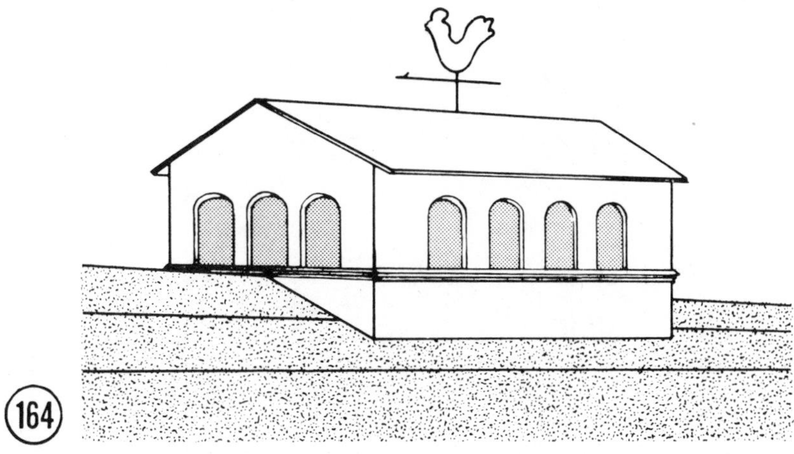

164

Those building a stable usually want to add that colonial touch only a cupola can provide, Illus. 164.

Illus. 165 shows a layout that simplifies cutting all parts from a ¾ x 4 x 8' panel of exterior plywood.

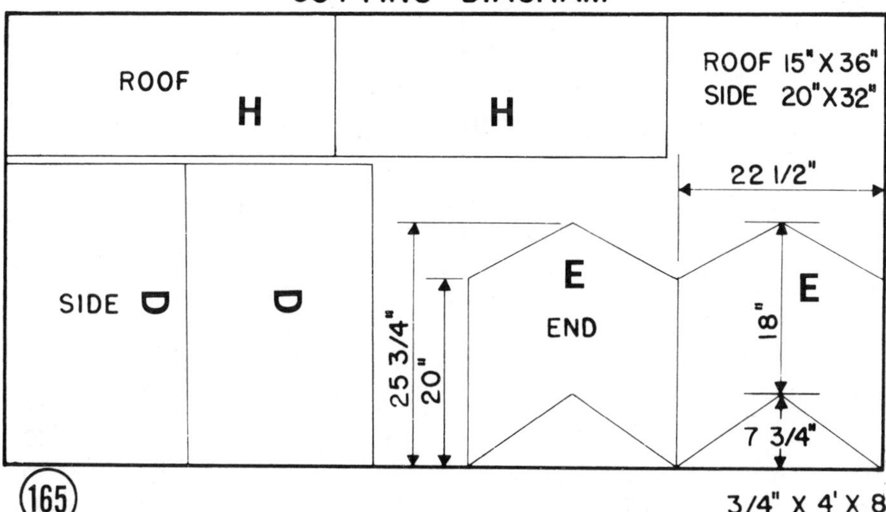

165

To fit a cupola to pitch of roof, do this. Place a scrap piece of 1 x 6 on the roof, Illus. 166. Hold a level vertically alongside. When bubble is centered, draw line on 1 x 6. Saw board along this line and use it as a pattern to cut angle on end of A, B, Illus. 167, 168.

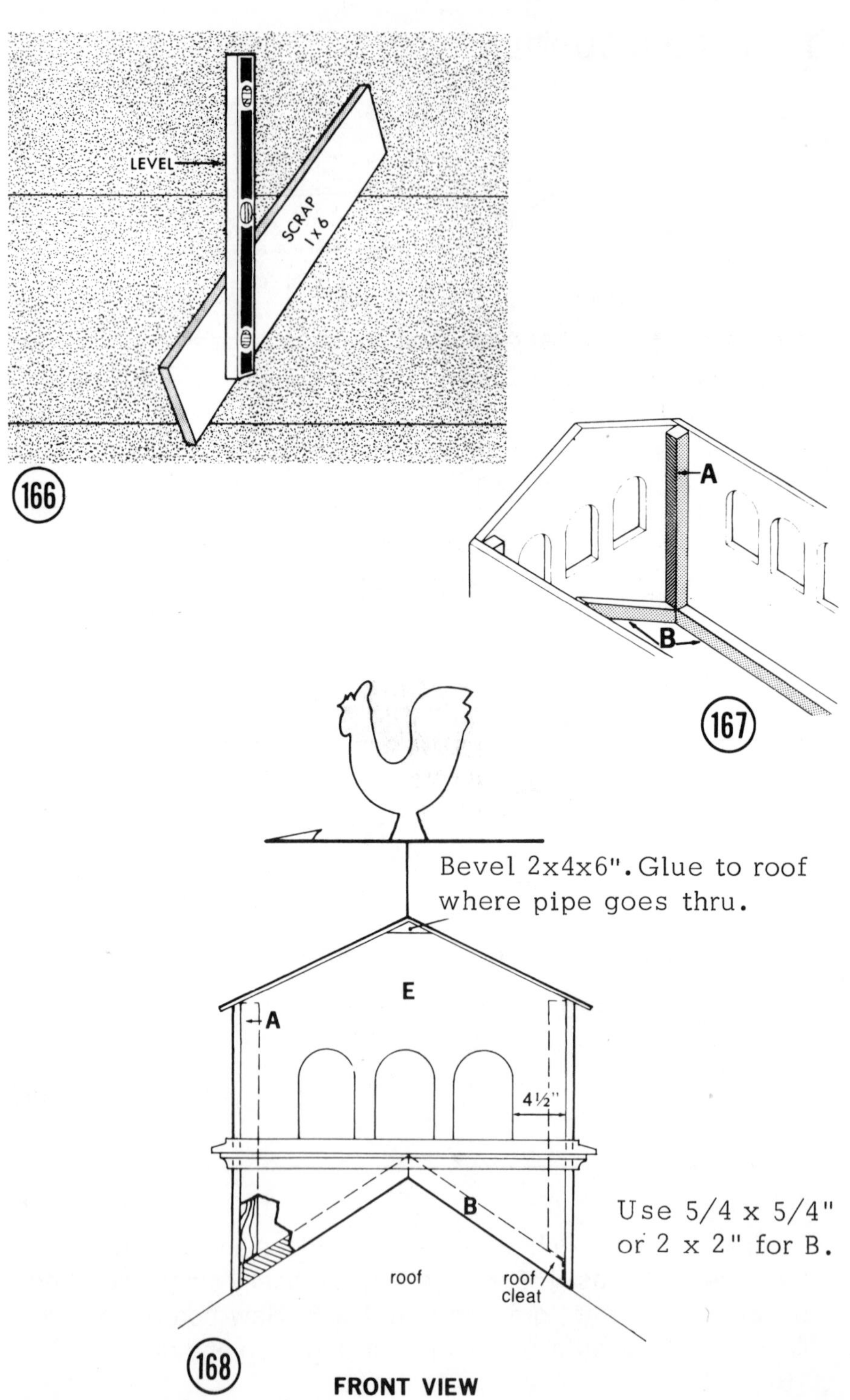
LEVEL
SCRAP
1x6
166
A
B
167
Bevel 2x4x6". Glue to roof where pipe goes thru.
E
A
4½"
B
roof
roof cleat
Use 5/4 x 5/4" or 2 x 2" for B.
168
FRONT VIEW

Cut two sides D, and two ends E, from ¾" exterior grade plywood, Illus. 165, 169, 170. Saw notch to fit roof. Plane bottom edge of D to shape of roof; plane top edge to fit roof of cupola.

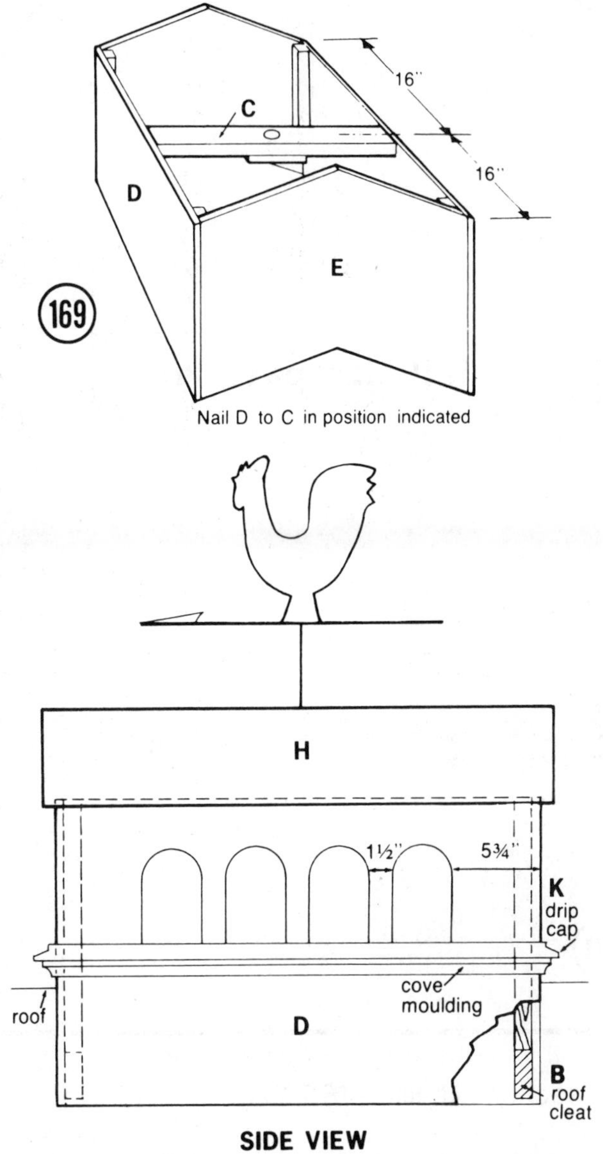

Nail D to C in position indicated

SIDE VIEW

Cut openings in D and E to size shown, Illus. 171. Cover openings on inside with screening if you use the cupola to house an alarm bell. Apply pieces of plywood on inside face, to seal openings, if you don't use screening.

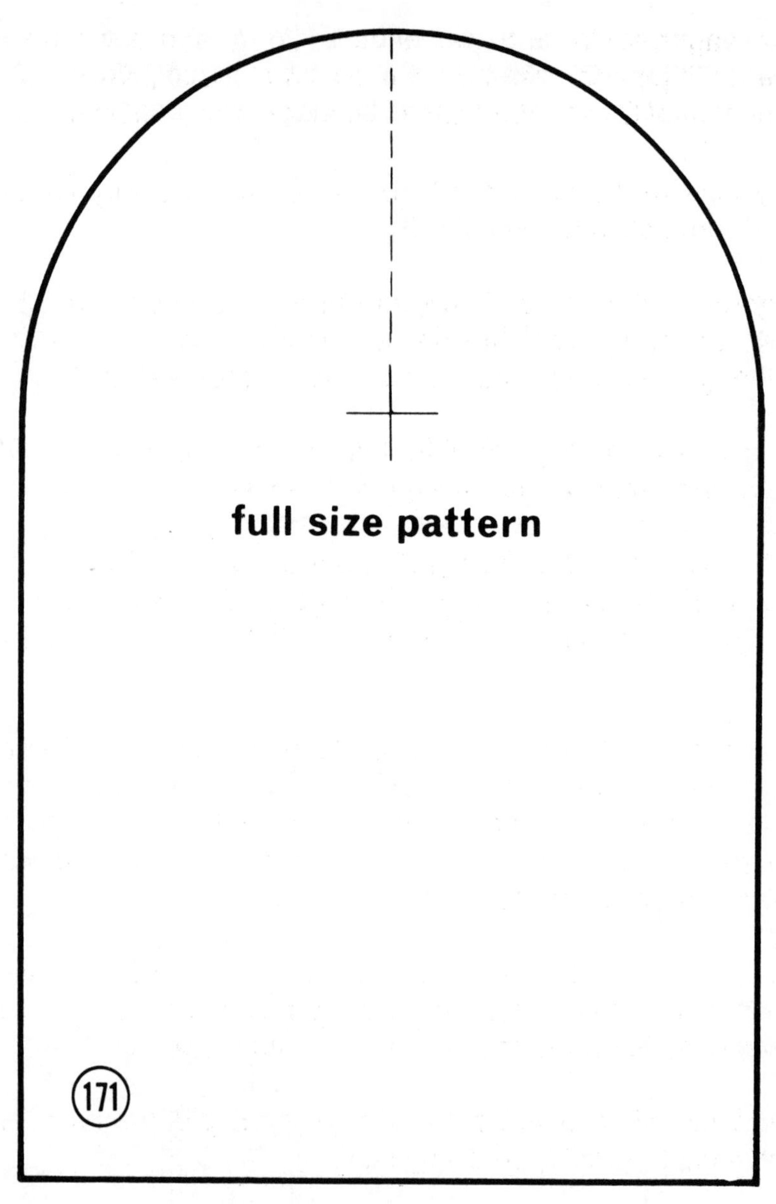

Cut two roof sections H, Illus. 165, 170.

If a weathervane is to be installed, nail 1 x 6 C in position shown. Easi-Bild Pattern #524 provides full size outlines for a weather-vane.

Glue and nail D to A, D to E and E to A with 6 penny finishing nails.

Draw diagonals from corner to corner of C and drill a hole at center to fit pipe for weathervane. Nail a piece of plywood or 1 x 4 to bottom of C to support end of weathervane pipe holder.

Apply glue and nail roof H to E and D with 8 penny finishing nails. H projects over end of E 2".

Cover roof with copper. Fold edges under roof overhang. Solder joints, or apply roll roofing to match roofing on roof. If weathervane is to be installed, cut hole to receive shaft.

Miter cut ends of drip cap K and cove molding, Illus. 170. Glue and nail in position all the way around cupola.

Cut four roof cleats B, Illus. 168, from 5/4 x 5/4. To determine length of B, measure from inside edge of E to center of ridge, Illus. 168. Cut ends of B same angle as shown, Illus. 166. Note position of angle cut.

Place cupola at center of ridge. Mark its location. Remove cupola and, allowing for ¾" thickness of E, fasten 2 x 2 cleats B, Illus. 168, to roof with 8 penny nails. Drill six holes through each E to permit fastening E to B with 2" No. 12 flathead wood screws—three screws to each B.

Paint inside and outside of cupola with at least two coats of outside paint. If you apply mineral surfaced roofing to cupola, allow roofing to project over edge of sheathing ⅜".

When you drill holes in E prior to fastening E to B, drill holes on a slight angle to simplify screwing into B.

If you install a weathervane, drill hole to size weathervane pipe holder requires. Insert pipe holder through hole and allow it to rest on C. Apply calking to seal pipe in roof.

After fastening ends E to B, insert weathervane in pipe. Countersink all screw heads, fill holes with putty. Touch up with paint.

The cupola makes an ideal place to install a bell or horn from a protective alarm system. Use screening over ports if you install a bell. Read Book #695 How to Install Protective Alarm Devices for information.

BUILD A MODEL

For building scale model of stable cut framing lumber to following size.

2 x 4 — ⅛ x ¼"	1 ft. = ¾"
2 x 6 — ⅛ x ⅜"	16" = 1"
2 x 8 — ⅛ x ½"	4' = 3½"

If you have never built a house or a barn, building a two story stable can look like a big deal. In reality, it only requires making one saw cut, or driving one nail at a time. Since it's a time consuming job that will require all your spare time, become familiar with every stage of construction. As previously mentioned, read the book through completely. Discuss any problems that may arise with your lumber retailer. If you still have any doubts, build a model. Using a scale of ¾" equals 1', you can build a 15 x 22½" model that makes a priceless heirloom for a child who likes horses. Taking care of a "stable" and its boarders, can teach a youngster much. Let the children help build and that, too, will widen their sphere of interest.

Your local hobby shop sells balsa wood and basswood in every needed thickness from 1/32" on up. They will be glad to suggest a scale for the 2 x 4, 2 x 6, 2 x 8 and 1 x 4 or 1 x 6 needed. Use small ½" brads and glue to assemble the model.

LIST OF MATERIALS
20 x 30′ Stable

Due to the variance in lumber width and thickness, the following is an approximate amount of material required to frame and sheath the stable. If you only build over weekends, and materials are left unguarded during the week, purchase material as needed. Purchase additional items as required. Plus 1 x 4 wind braces.

FRAMING

LEFT SIDE
Shoe — 2 x 4 — 2/16′
Plates — 2/8′, 2/12′, 2/14′
Studs, Headers, Sills — 30/8′

RIGHT SIDE
Shoe — 2 x 4 — 2/16′ Headers — 2 x 6 — 3/10′
Plates — 2/8′, 2/12′, 2/14′
Studs — 18/8′

FRONT
Shoe — 2 x 4 — 1/8′, 1/12′ Headers — 2 x 6 — 1/10′
Plates — 2 x 4 — 1/8′, 2/10′, 1/12′
Studs, Sill — 2 x 4 — 18/8′

REAR
Shoe — 2 x 4 — 1/8′, 1/12′
Plates — 2 x 4 — 1/8′, 2/10′, 1/12′
Studs, Header, Sill — 2 x 4 — 18/8′

JOISTS, BRIDGING — 2 x 8 — 19/20′ or 19/10′ and 19/12′

LOFT FLOORING — 19 — 4 x 8 x 5/8″ or ¾″ plyscord to cover 600 sq. ft.

SHEATHING
Side Walls, Gable Ends — 35 — 4 x 8
Use ½″, ⅝″ or ¾″ exterior grade plywood
Roof — 32 — 4 x 8 — 1/2″, 5/8″ or ¾″ plyscord

RIDGE — 2 x 8 — 2/16′
1 — Truss Clip — 3-7/16 x 12½
1 — 2¾ x 10 splice clip

RAFTERS — 2 x 6 — 32/16′ for A, B
2 x 4 — 15/12′ for C
2 x 4 — 14/8′ for D
64 — truss clips — 3-7/16 x 12½″

GABLE STUDS — 2 x 4 — 5/10′, 7/12′, 2/14′, 4/16′

ROOFING — 110 to 120 Double coverage, mineral surfaced roll roofing, or galvanized steel or aluminum roofing to cover 960 square feet. Buy nails and asphalt cement or sealer manufacturer recommends.

AISLE WALL
2 — 2 x 8 x 16′ — A
6 — 2 x 6 x 8′ — B
3 — 2 x 6 x 10′ — C
2 x 8 — 8/8′, 4/10′, 4/12′
6 — truss clips — 3⅜ x 7½
6 — 4″ x 8′0″ x ⅛″ or 3/32″ steel or aluminum for gusset plates

DOOR FRAMES — 2 x 6 — 4/16′, 2/10′

DOOR CASINGS — 1 x 4 — 4/16′, 2/10′

DOORS — 3 — Dutch Doors
2 x 8 — 18/8′
2 x 4 — 3/8′
1 x 6 or 5/4 x 6 — 6/8′, 3/12′ No. 2 Clear Pine
3 — Sliding Doors
3 — ¾ x 4 x 8 Exterior Grade Plywood
or 18 — 2 x 8 x 8′ plus 3 — 2 x 4 x 8′
1 x 6 or 5/4 x 6 — 9/8′, 3/12 No. 2 Clear Pine
1 — Entry Door
1 — ¾ x 4 x 8 Exterior Grade plywood
5 — 1 x 6 or 5/4 x 6 x 8′ No. 2 Clear Pine

STALL LINER

Use 2 x 8 free of knots
2 x 8 — 34/10', 7/14', 7/16'
7 — 3/4 x 4 x 8 plyscord
10 — 1 x 2 x 16'

AISLE WALL AND DIVIDERS

8 — 1½ x 1½ x 8'0" angle iron
1½ x 1½ x 1½ — U channel (optional) — 40 lineal feet.

WINDOWS, WINDOW FRAMES, CASINGS

5 — 2'4" x 1'6" — 3 lite cellar sash or purchase size available
1 — 2 x 8 x 16' Sill
3 — 1 x 6 x 10' No. 2 Pine — Frame
3 — 1 x 4 x 10' No. 2 Pine — Casings

WINDOW GUARDS

3 — Iron Window Guards or
To build guards you will need:
1 — 1 x 10 x 8' Sides
4 — 1 x 2 x 10' Rails

SLIDING DOOR HARDWARE

The hardware illustrated is readily available at leading hardware, building material and farm supply stores. Each sliding door requires:

1 — 8'0" #51 Trolley Rail
3 — #51F — Single Trolley Rail Brackets
2 — #51 End Caps
2 — Stay Rollers #318 or #319
2 — Door Bumpers #19
1 — #13 Door Latch

Each outside Dutch Door requires:

2 pairs Extra Heavy 6" or 8" T Hinges
plus 36 — 1¼" #12 flathead screws
1 pair #14/15 Bar Holders
2 #13 or #13B — Door or Gate Latch

Entry Door — 1 — ¾ x 4 x 8 Exterior Grade Plywood
5 — 1 x 6 or 5/4 x 6 x 8' No. 2 Clear Pine

FASCIA — 1 x 6 — 4/16' No. 2 Clear Pine
1 x 4 — 4/16', 4/10' No. 2 Clear Pine

BATTENS — ¾ x 2½" — 77/8, 12/10

GRILLE — ½" aluminum or galvanized pipe

DRIP CAP — 2/16'

CUPOLA — 1 — ¾ x 4 x 8' Exterior Grade Plywood
1 — 2 x 4 x 10' or 2 x 2 x 10'
Drip Cap — 1⅙ x 1⅝" — 10/12'
¾ Cove Molding — 10/12'

MATERIALS—FOOTINGS, FOUNDATION

Exact amounts of cement, sand, gravel or total number of concrete blocks depend on depth and thickness of footings, and height of foundation wall required by local building codes. As a starter, order:

10 bags of cement
5 yds. of sand
5 yds. of ¾" or run-of-bank gravel
½" x 12" machine bolts, nuts, washers

Approx. 74-6" x 8" x 16" concrete blocks per course below grade (due to breakage, order a few more blocks than actually required).

RED BARN TOOL HOUSE

(172)

A tool house measuring 8′0″ x 10′0″, Illus. 172, can be built by following the same general procedure.

Build it on a concrete slab, Illus. 173. A 16″ x 24″ deep trench around perimeter, Illus. 174, is ample for most soil conditions. Follow building codes concerning depth below frost level.

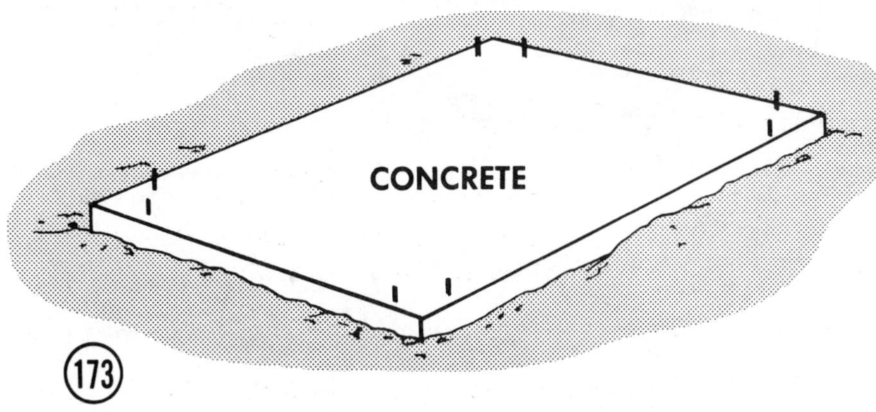

(173)

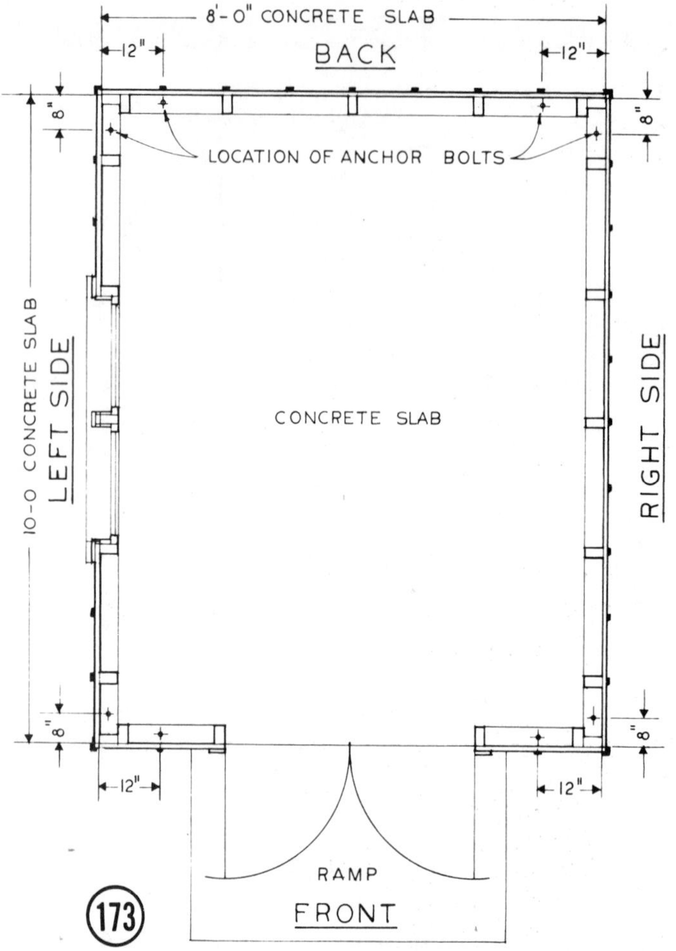

8'-0" CONCRETE SLAB
12"
BACK
12"
8"
8"
LOCATION OF ANCHOR BOLTS
10-0 CONCRETE SLAB
LEFT SIDE
CONCRETE SLAB
RIGHT SIDE
8"
8"
12"
12"
RAMP
FRONT

173

FORM
CONCRETE
REINFORCING WIRE
GRADE
GRAVEL
CONCRETE
FIELDSTONE

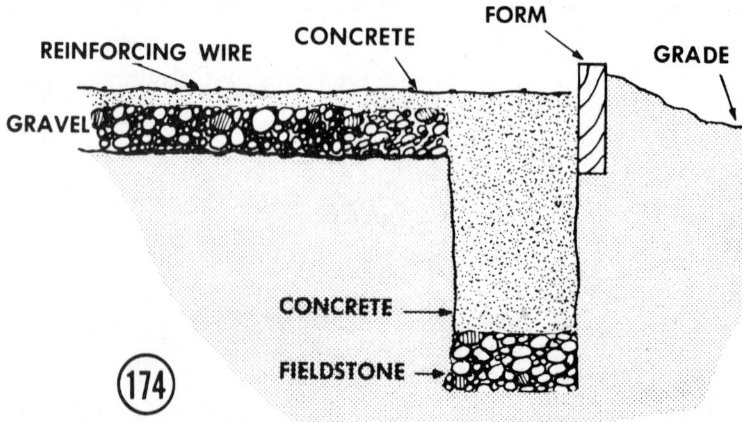

174

Set up a 2 x 8 form, Illus. 175, with inside dimensions 8′0″ x 10′0″. Hold square with temporary braces across corners. Measure diagonals. Form is considered square when diagonals are equal length.

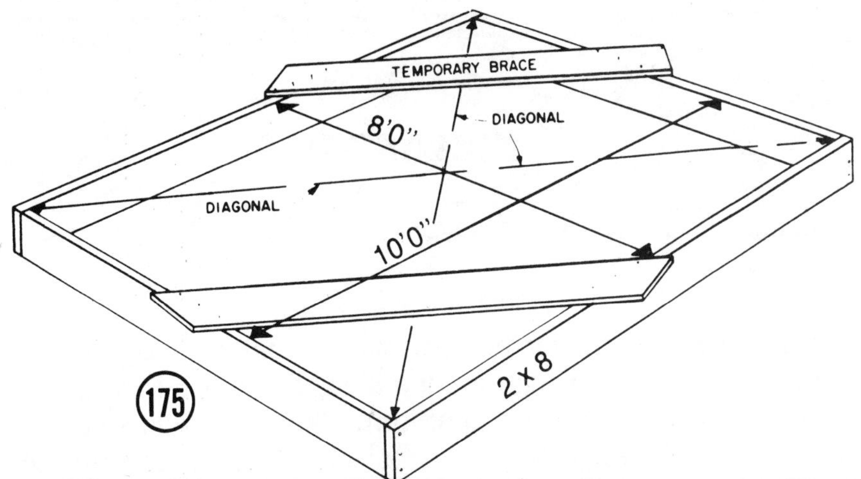

To visualize size, and its relation to a house, garage and distance from property line, place form on site selected, parallel to, or at right angle to other buildings.

Dig a 16″ trench, 24″ or to depth codes recommend. If you have fieldstone, partially fill trench.

Place form around excavated area, keeping top edge approximately 7½″ above highest point of adjacent ground. Check with level. Shim with stones to level. Paint inside of form with old crankcase oil. The top edge of form now represents top of finished slab. Backfill around form to hold in position, Illus. 176.

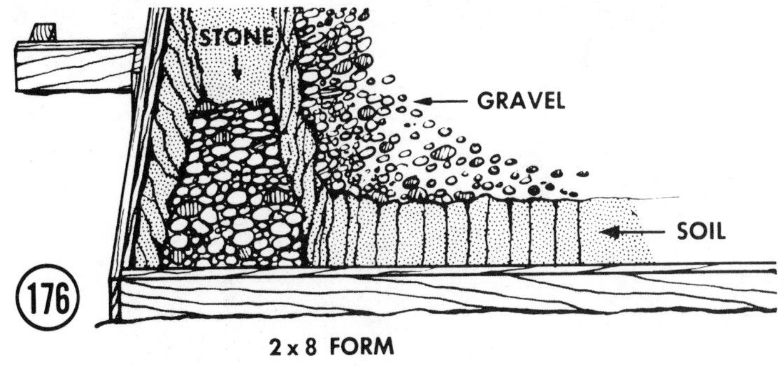

Excavate floor area to permit laying about 3″ of gravel, 1½″ of concrete, Illus. 177.

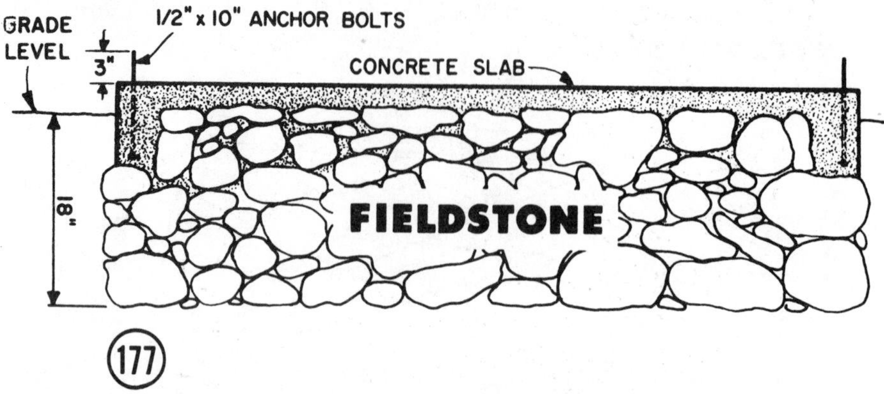

177

Position ½ x 10″ anchor bolts where indicated, Illus. 173. To hold bolts in position while pouring slab, drill holes in scraps of 2 x 4 and tack these to edge of form, Illus. 178. Allow bolts to project 3″ above edge of form. Position center of bolt an 1½″ from inside face of form. Pour slab using one part cement, three parts sand to five parts gravel, or buy readymix. Pour concrete to about an inch from top of form and embed 6 x 6 reinforcing wire over entire area. Cover with concrete and level floor with a 2 x 4 screed. You can float surface smooth using a steel or wood float, Illus. 179. Allow slab to set at least three days, then remove form. Lay gravel, build form, pour ramp, Illus. 180.

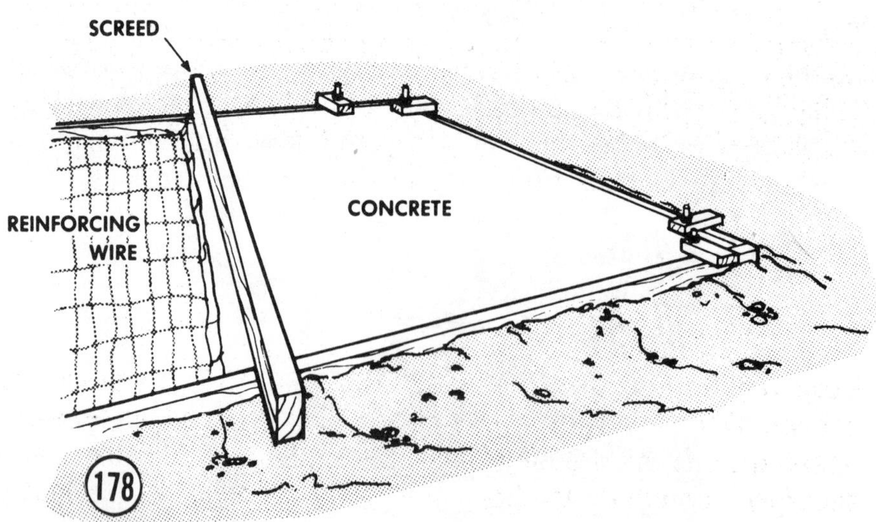

178

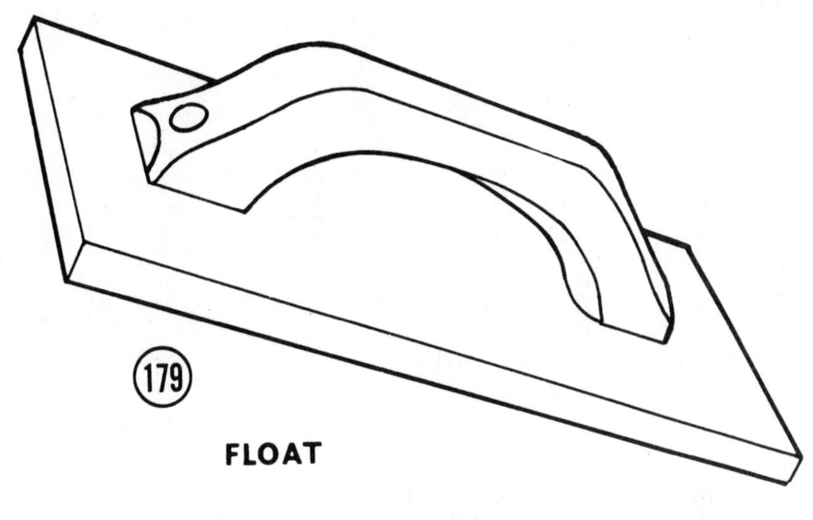

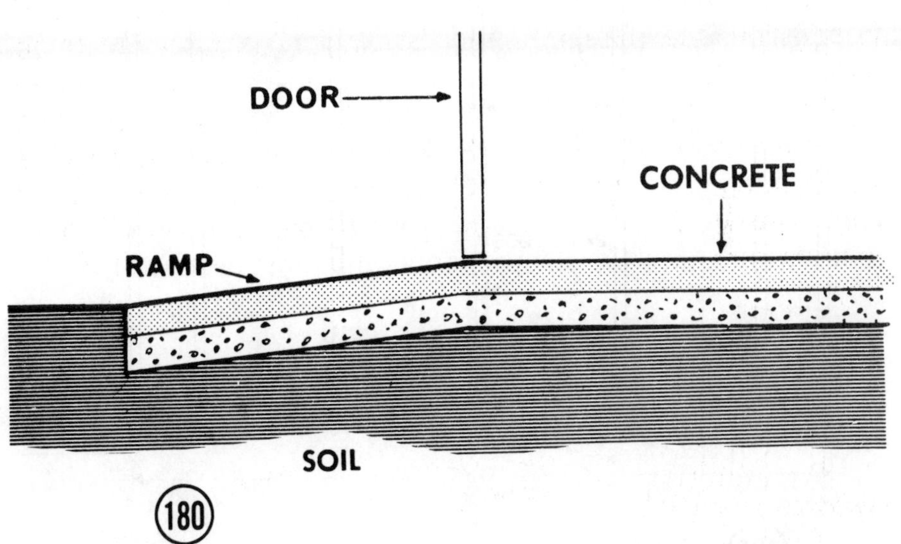

Build wall frames, Illus. 181, 182, 183, 184, 185, to overall size shown. Drill holes for anchor bolts. Don't cut shoe for door opening until all framing has been plumbed, leveled, and securely fastened to bolts following directions previously outlined.

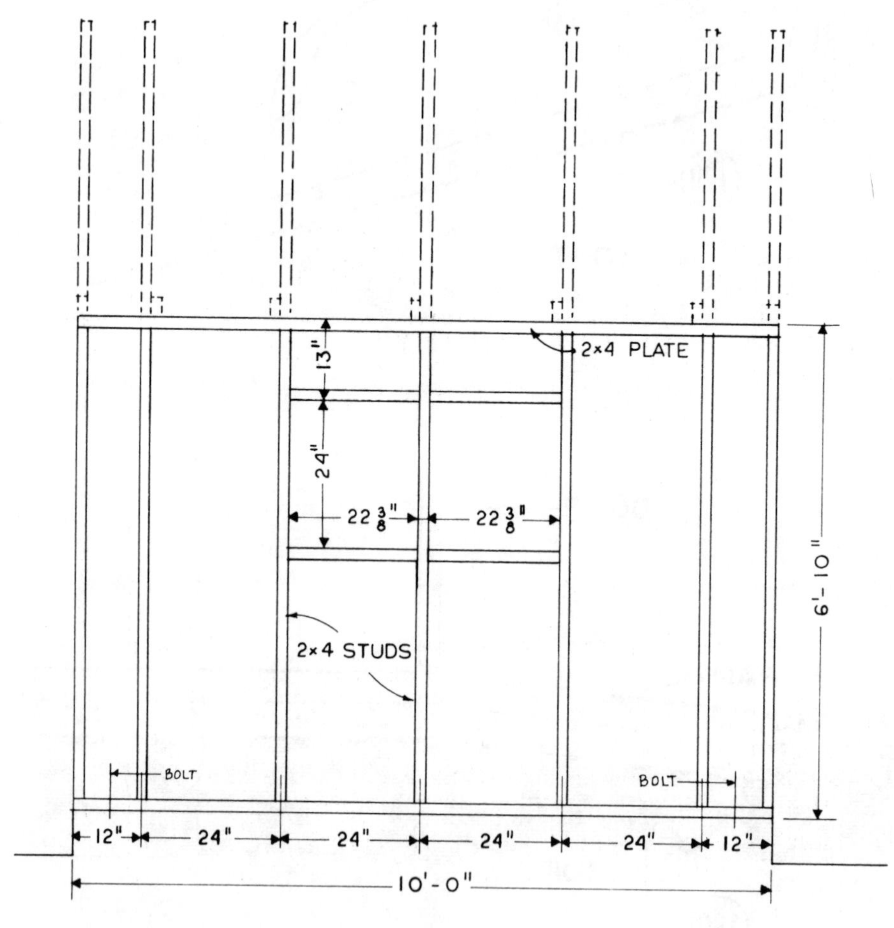

(181) **RIGHT SIDE FRAME**

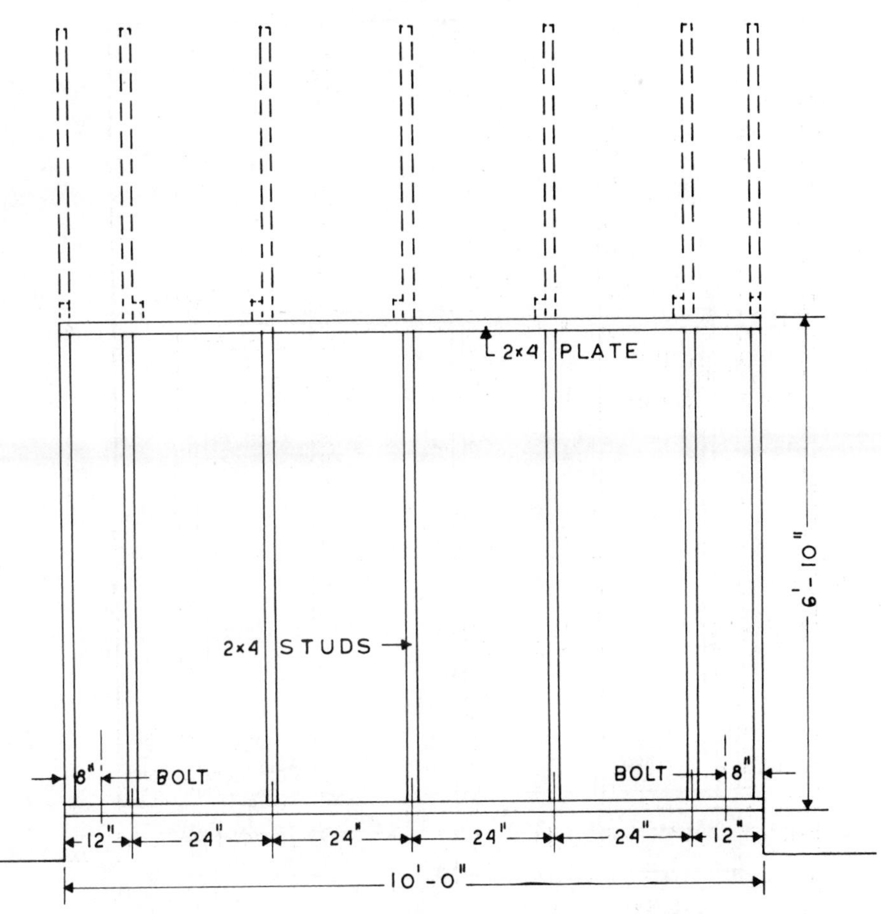

(182) **LEFT SIDE FRAME**

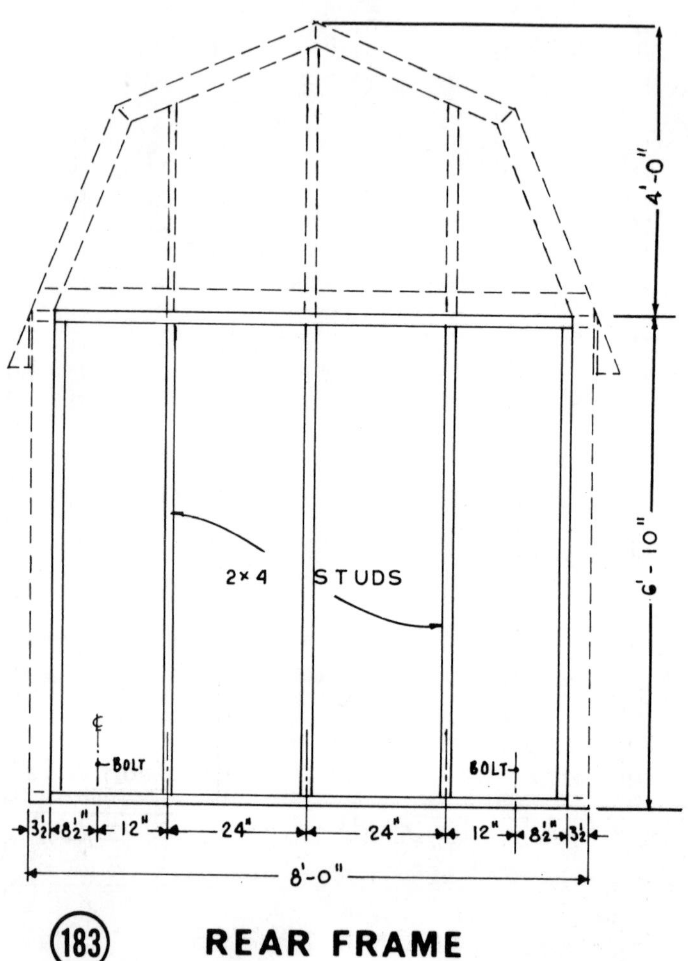

(183) REAR FRAME

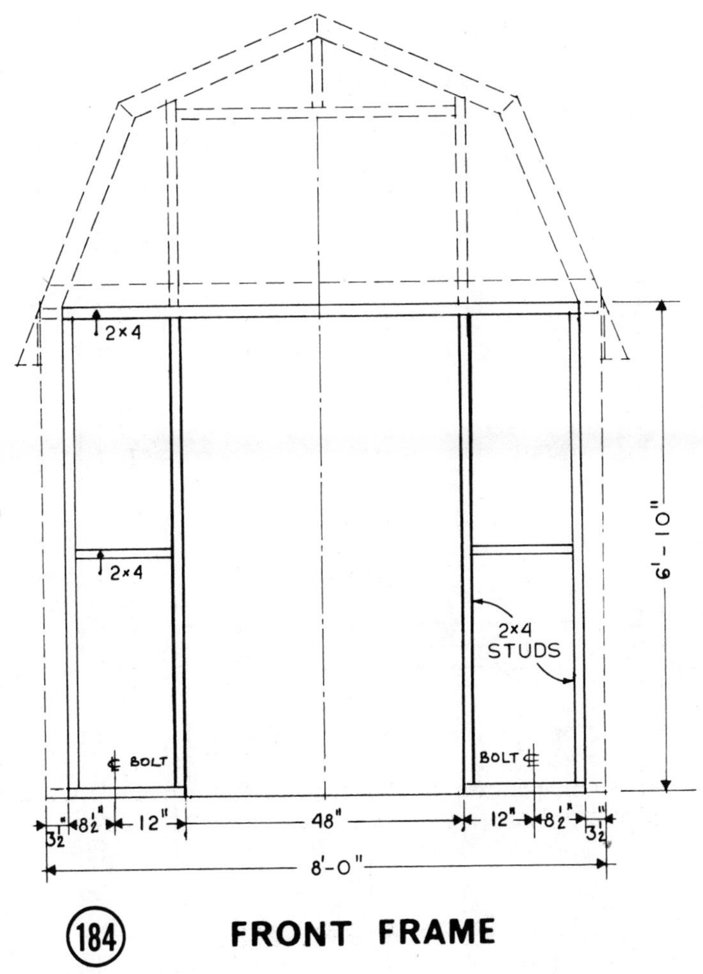

(184) **FRONT FRAME**

Nail 2 x 4 ceiling joists 2′0″ on centers in position indicated, Illus. 61.

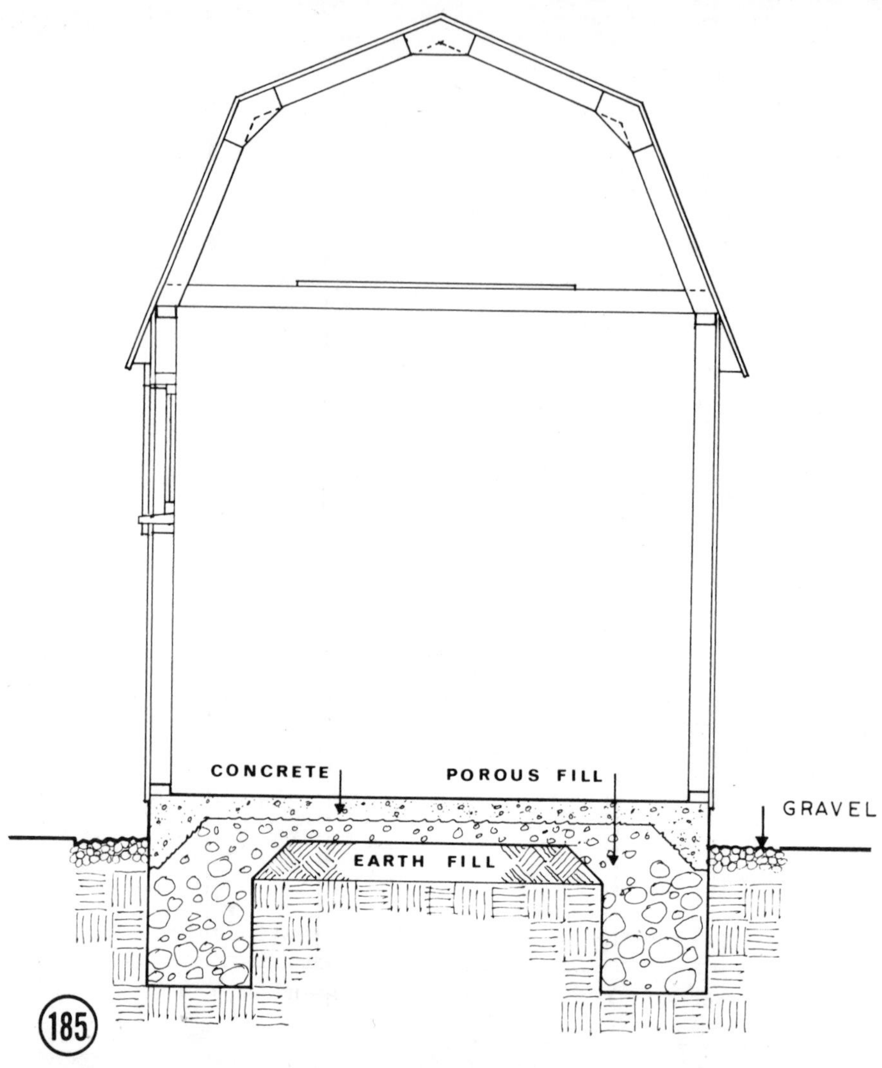

Illus. 186, provides full size angle pattern for rafter. To build a truss rafter, nail ½, ⅝ or ¾ gusset plates, Illus. 187, to both sides of rafter, instead of using 2 x 4C, Illus. 75. To insure cutting rafters to length your construction requires, cut a test pair about ⅜″ longer than required. Place in position. If O.K. recut to angle and length required. Be sure rafter A sets squarely on plate. When test rafters fit properly, fabricate truss rafters. Assemble and brace rafters following rafter construction previously outlined. Nail or bolt truss rafters to ceiling joists.

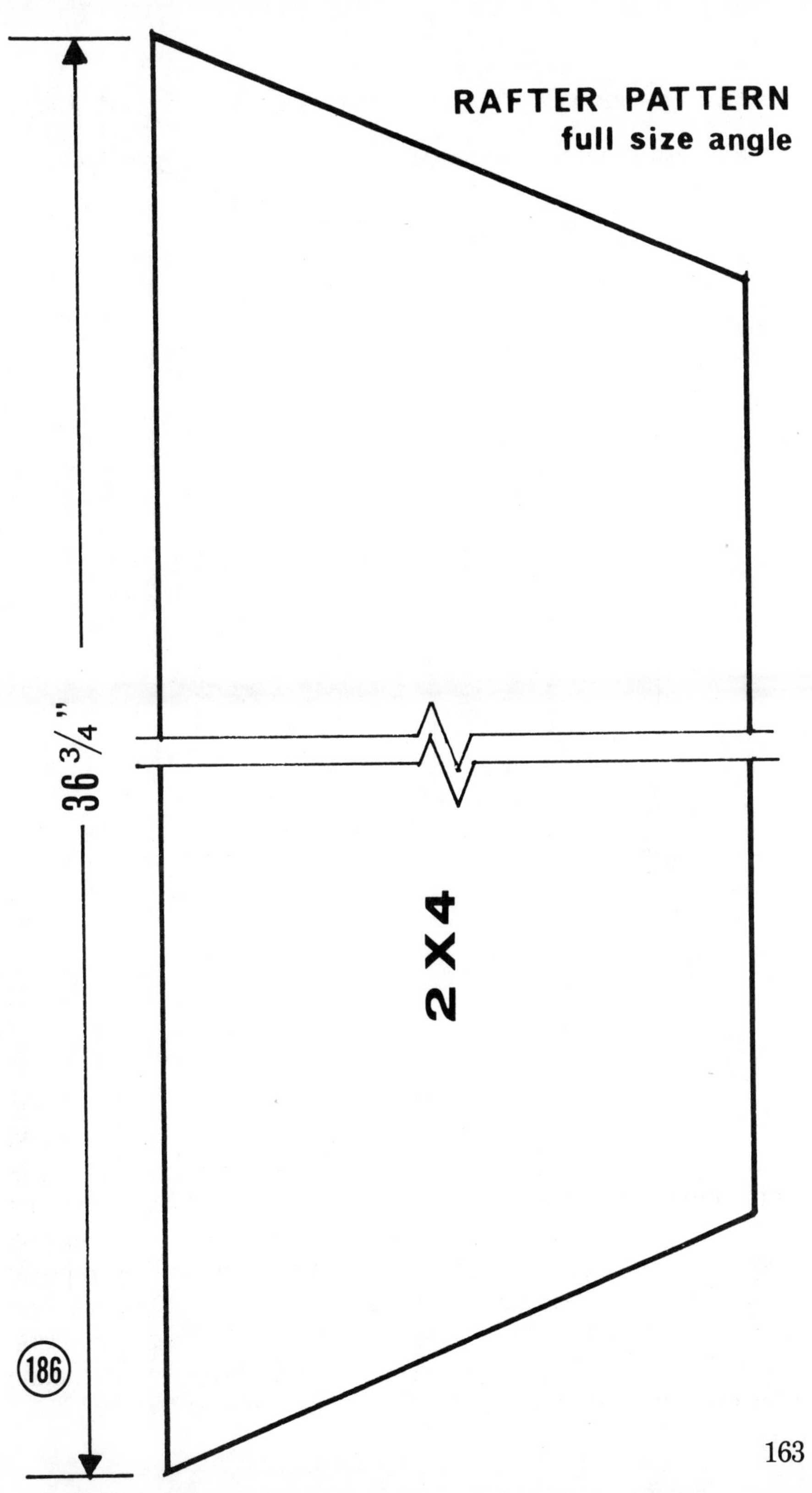
RAFTER PATTERN
full size angle
36 3/4"
2 X 4
186

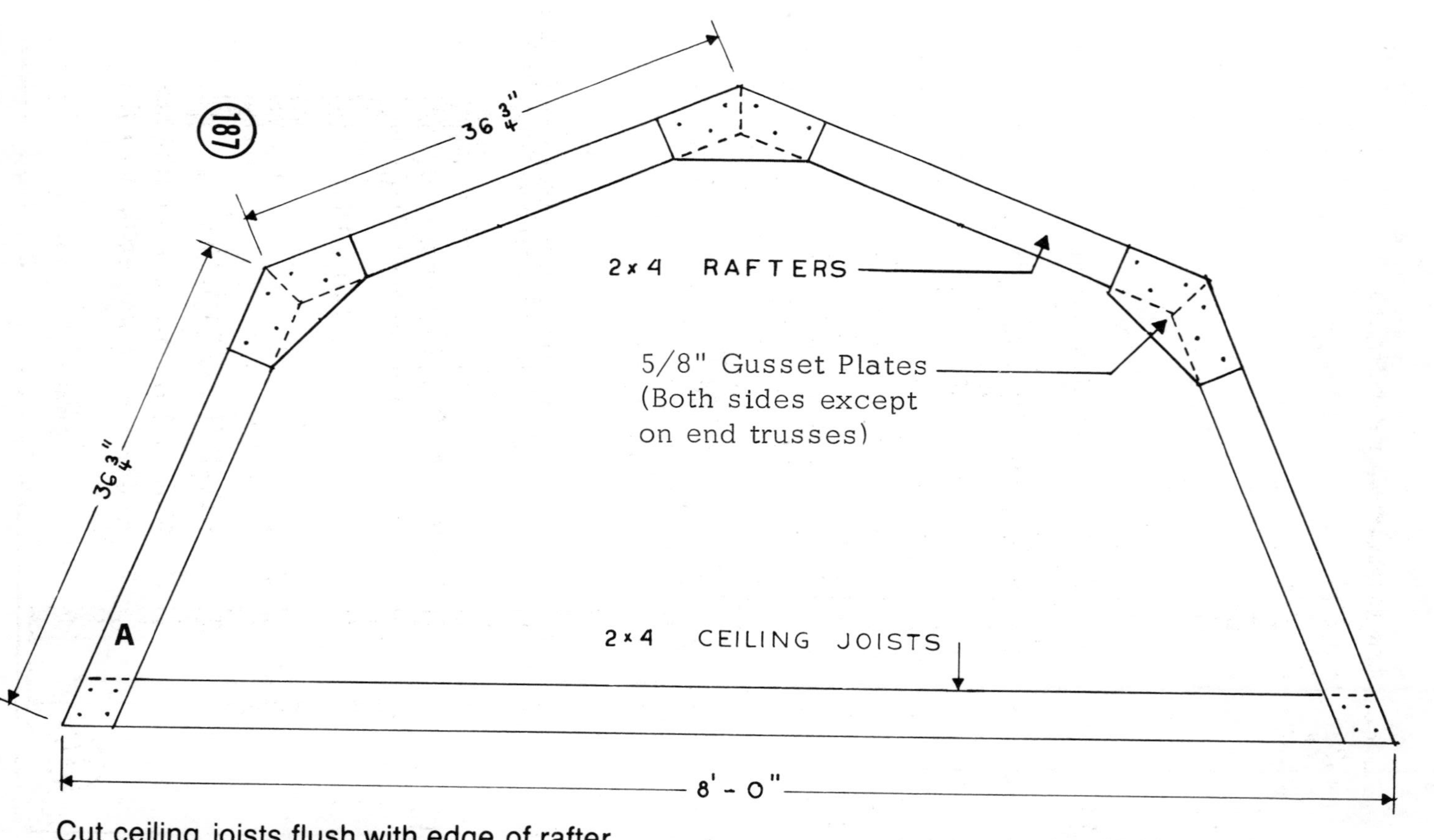

Cut ceiling joists flush with edge of rafter.

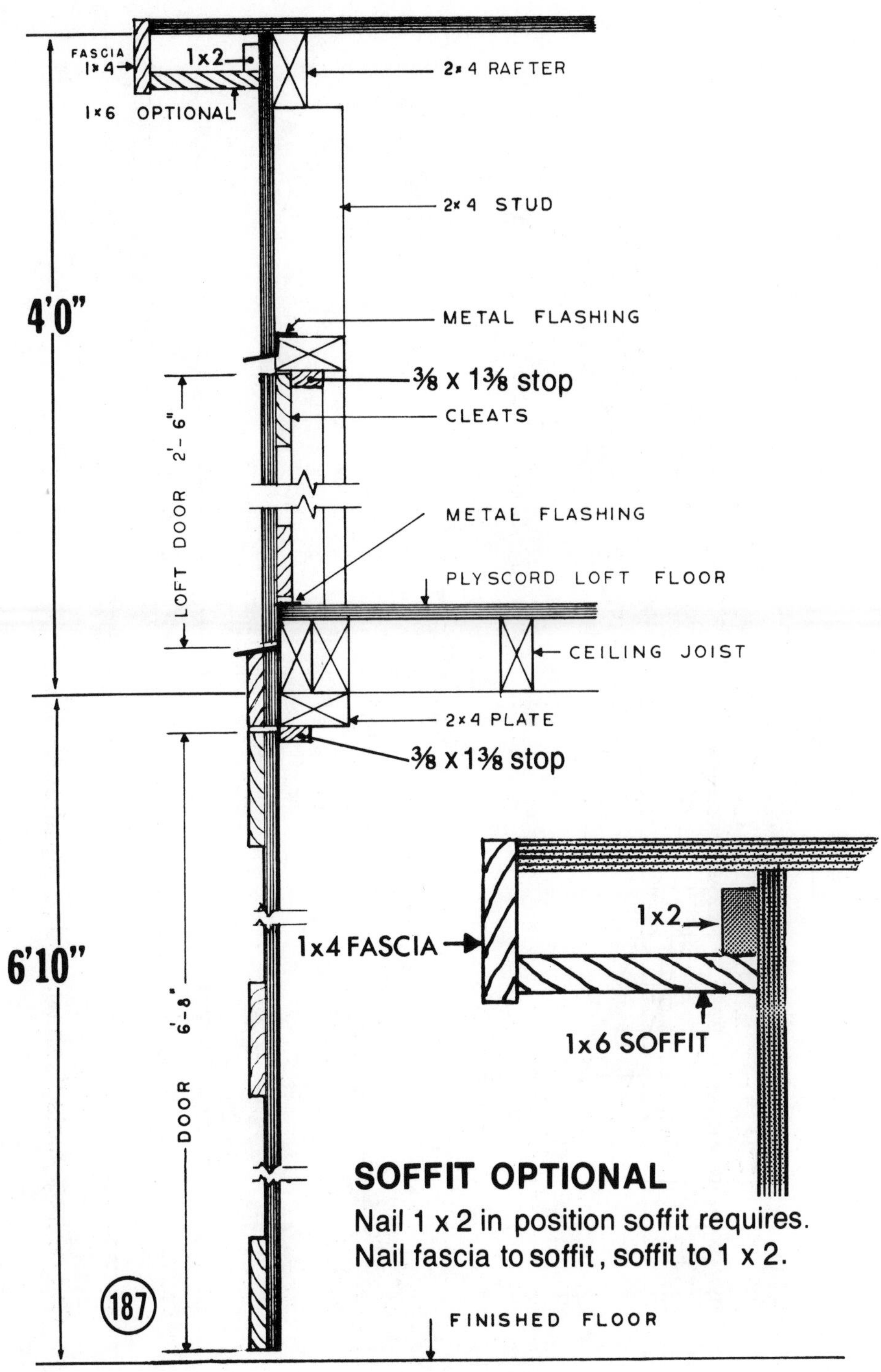
FASCIA
1x4
1x2
2x4 RAFTER
1x6 OPTIONAL
2x4 STUD
4'0"
METAL FLASHING
⅜ x 1⅜ stop
CLEATS
LOFT DOOR 2'-6"
METAL FLASHING
PLYSCORD LOFT FLOOR
CEILING JOIST
2x4 PLATE
⅜ x 1⅜ stop
1x2
1x4 FASCIA
6'10"
1x6 SOFFIT
DOOR 6'-8"
SOFFIT OPTIONAL
Nail 1 x 2 in position soffit requires.
Nail fascia to soffit, soffit to 1 x 2.
187
FINISHED FLOOR

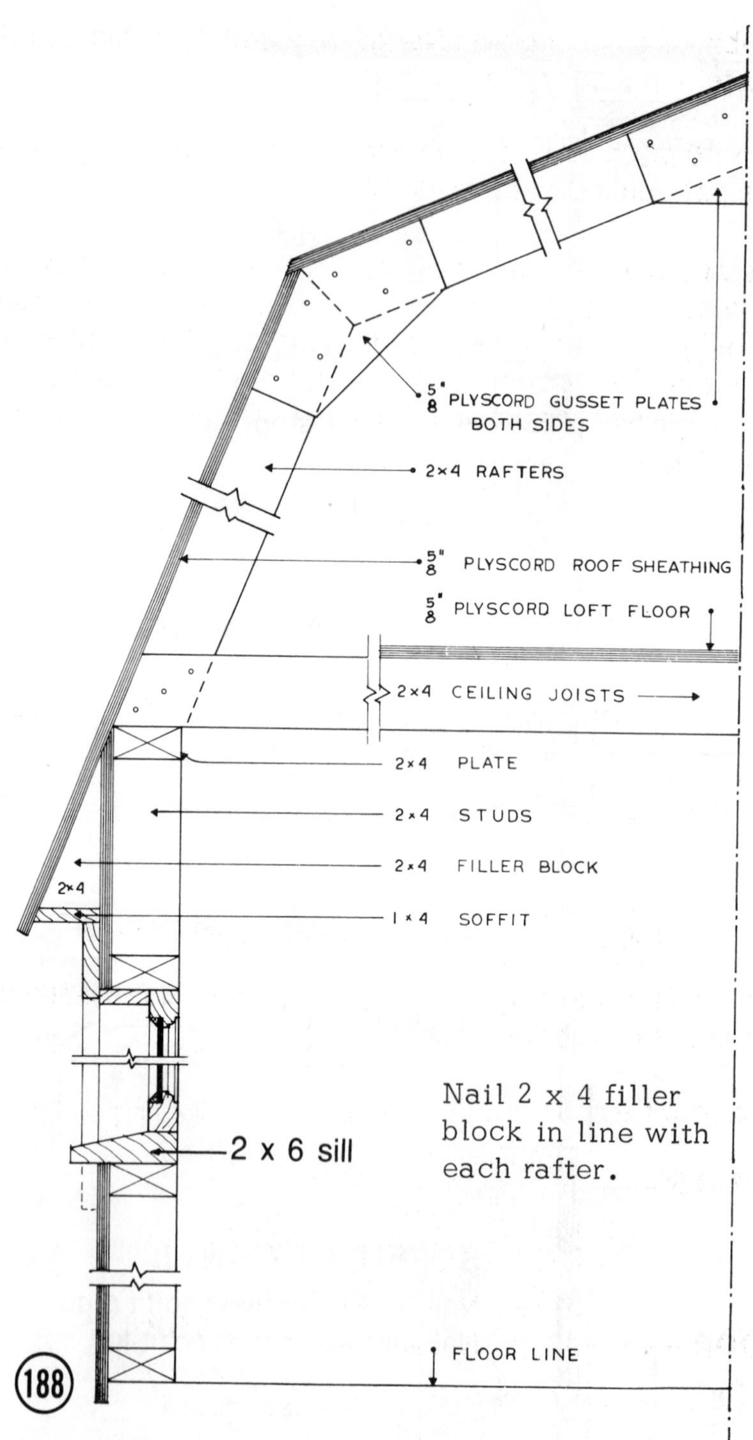

Nail 2 x 4 filler block in line with each rafter.

Cut gable studs to angle and length required. Nail studs to rafter and plate with 8-penny common nails.

⅝ x 4 x 10 plyscord is recommended for roof sheathing. Project panels position indicated, Illus. 188.

Before applying roof sheathing, cut 2 x 4 filler blocks, Illus. 189. Nail these in position shown. Nail first panel so it projects below edge of filler block. Nail through panel into rafters and filler blocks. Bevel edge of 1 x 4 and nail to bottom of filler blocks. This finishes flush with front and back filler blocks. Apply roof sheathing.

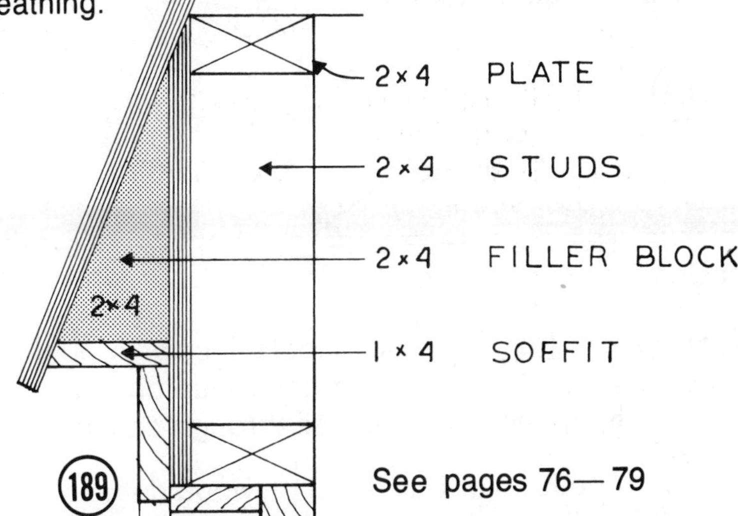

See pages 76—79

Cut 1 x 4 fascia to angle used for rafter. Nail roof sheathing to fascia in position shown, Illus. 187.

Cut shoe off across door opening.

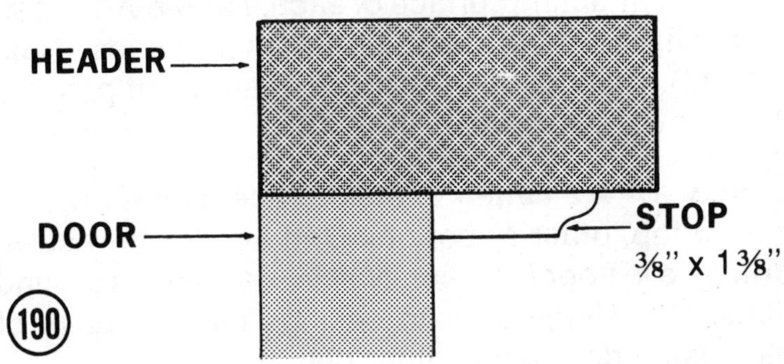

Build a plywood door using ¾″ exterior plywood and 1 x 4 framing to fit size of opening, Illus. 191, following directions previously outlined.

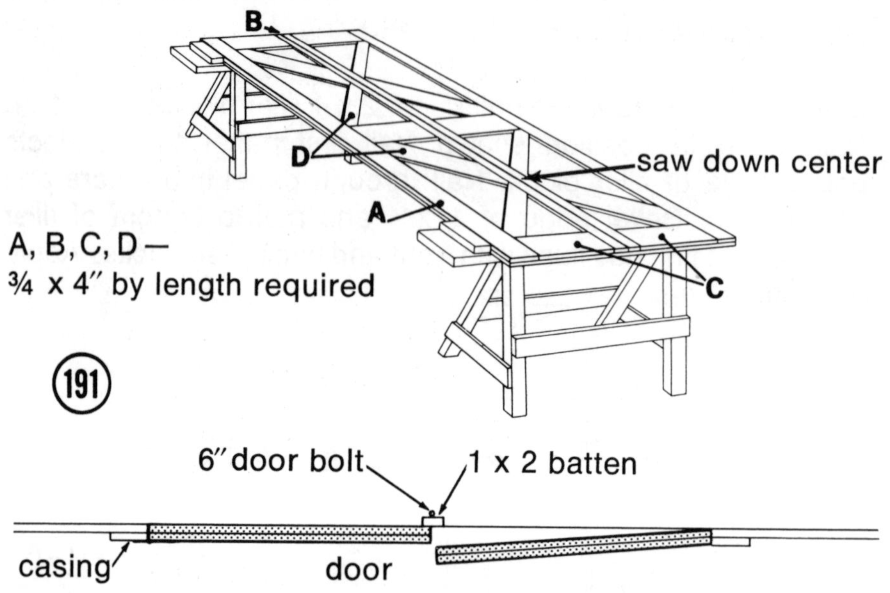

Door and window frames for a tool house are optional. Follow directions noted for stable or double up on studs alongside door opening and screw hinges through casings.

Follow procedure for building a plywood door as outlined on page 113, with this exception. Cut parts for rails, stiles and cross bars from ¾″ exterior grade plywood. Use a 4′0″ x 8′0″ ¾ exterior grade plywood panel for door. Cut two A, one B, six C, eight D— ¾ x 4″ by length required. Apply waterproof glue over entire contacting surface of each. Fasten A in position with 4 penny finishing nails. Countersink heads. Next apply B, C and D. Allow glue to set permanently. Fit panel to opening, then saw down center. Illus. 191.

Glue and screw a 1 x 2 batten to inside face. Fasten two 6″ door bolts, one top, other at bottom. Drill header to receive one bolt, drill hole in floor for other. Nail ¾ x 4 casings around opening, Illus. 114. Hinge door to casings using 3 pairs of 4″ T hinges. Fasten door latch, Illus. 124.

Windows can be hinged at bottom, Illus. 192, 193, to open in, with 2″ loose pin butt hinges. Nail stop in position indicated. Chains, fastened where indicated, hold window open amount desired. Window can be locked with 2″ barrel bolt.

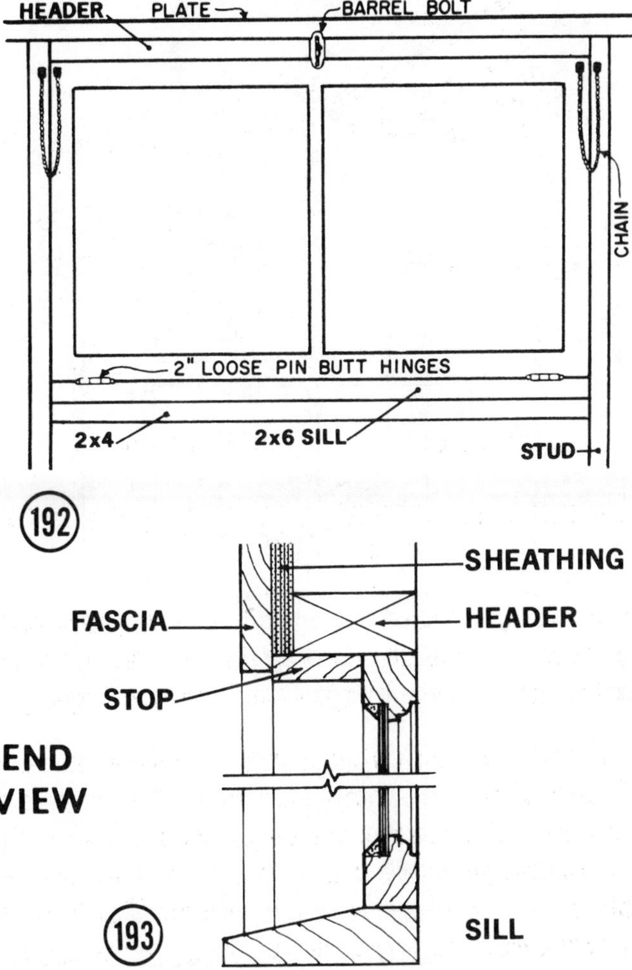

Use ⅝″ or ¾″ exterior grade plywood for siding. Nail to studs with 8-penny common nails. Apply battens as noted on page 112.

4″ aluminum louvers, installed in siding in both ends helps provide necessary ventilation.

Apply roofing following directions previously outlined.

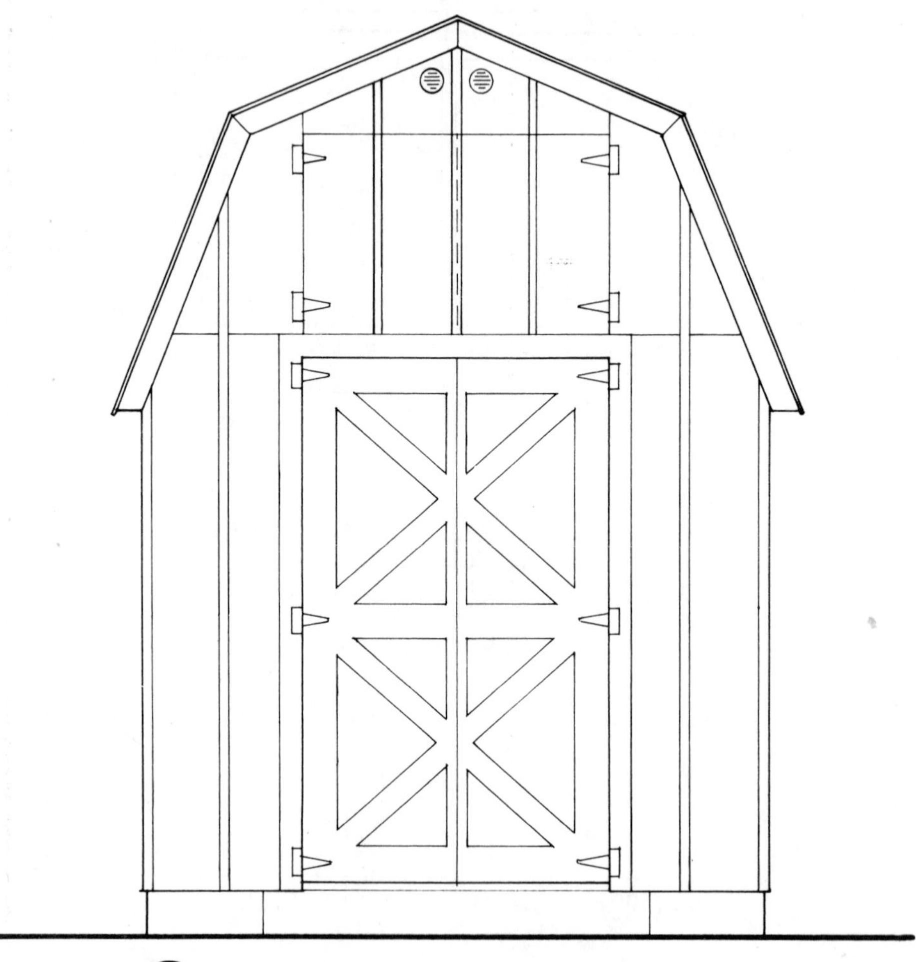

(194) FRONT ELEVATION

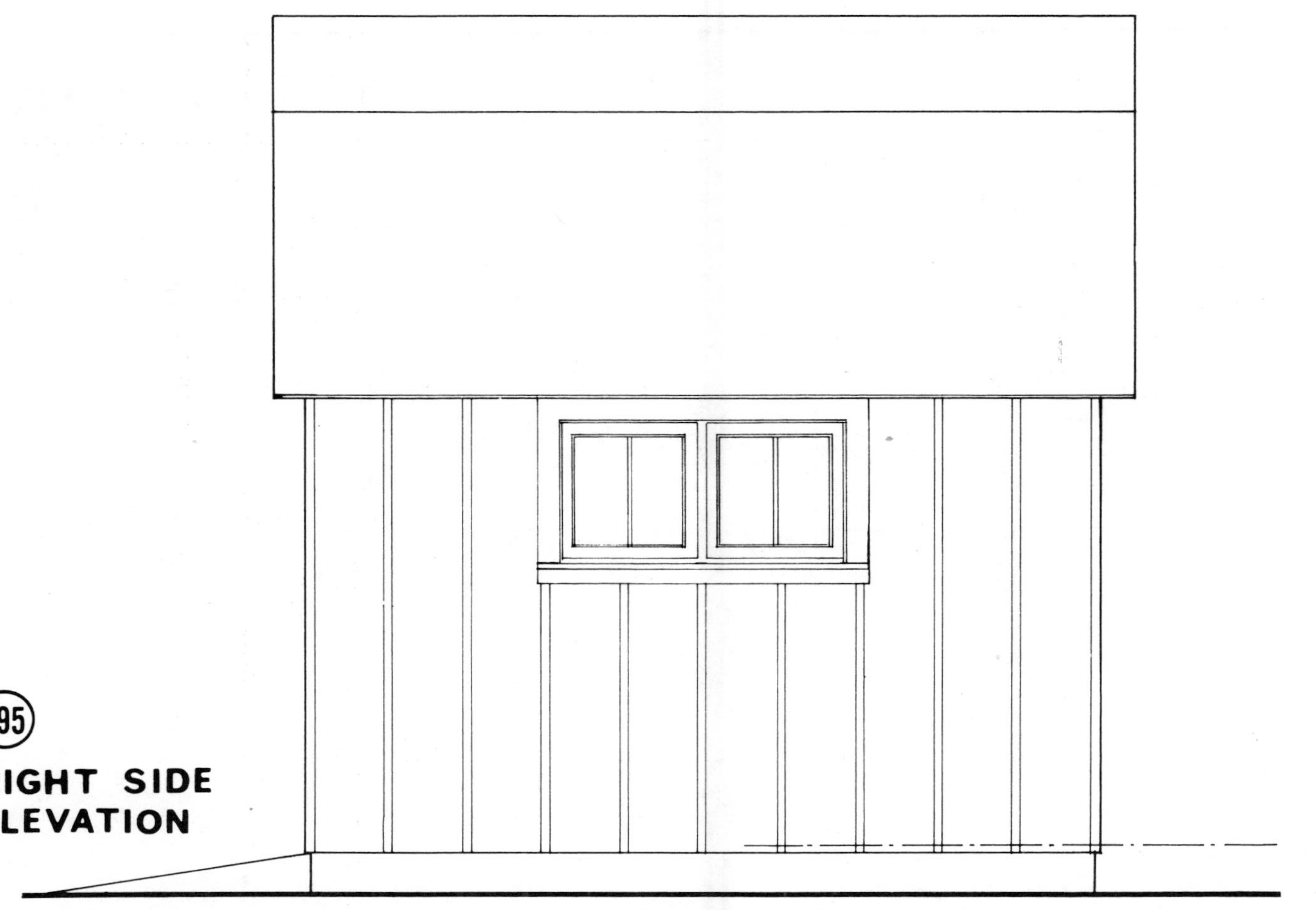
195
RIGHT SIDE
ELEVATION

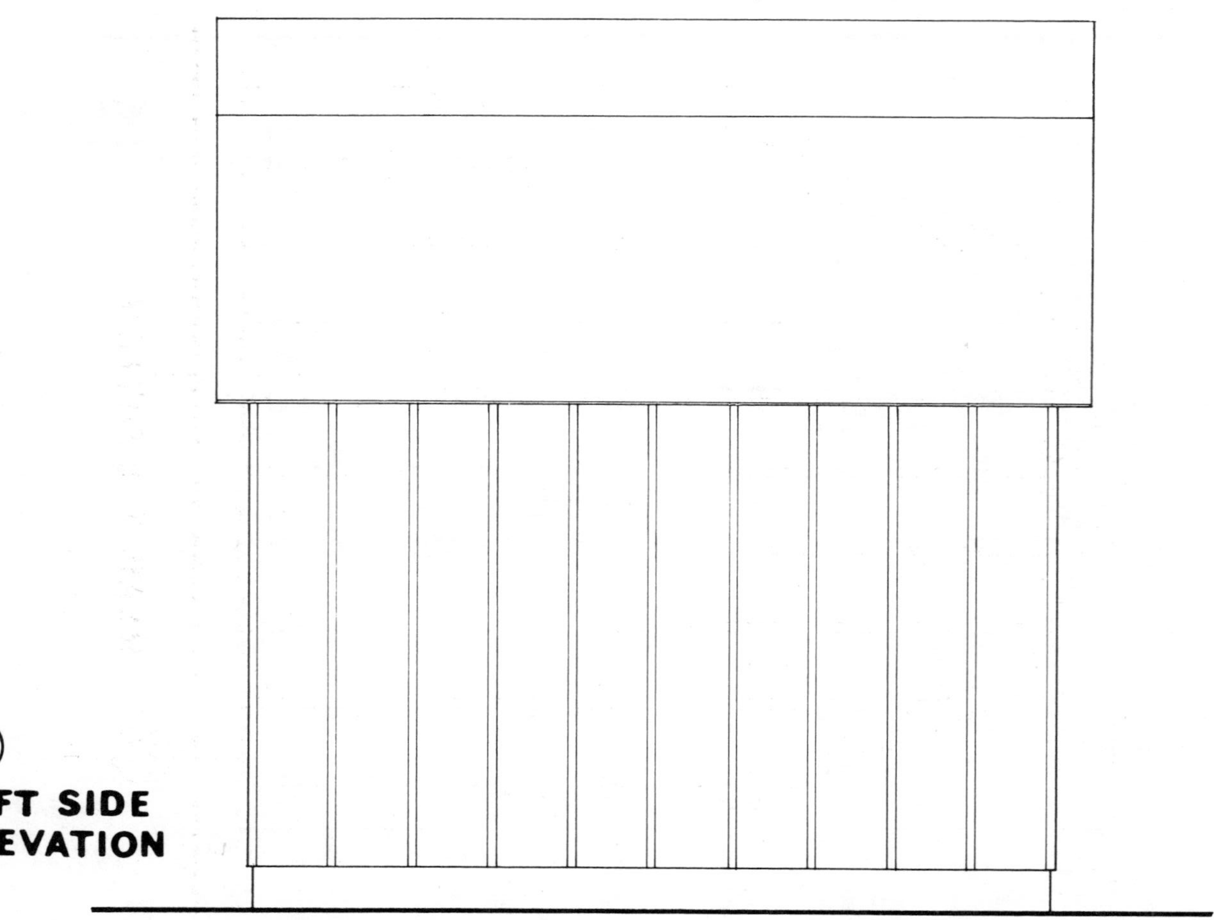

196 LEFT SIDE ELEVATION

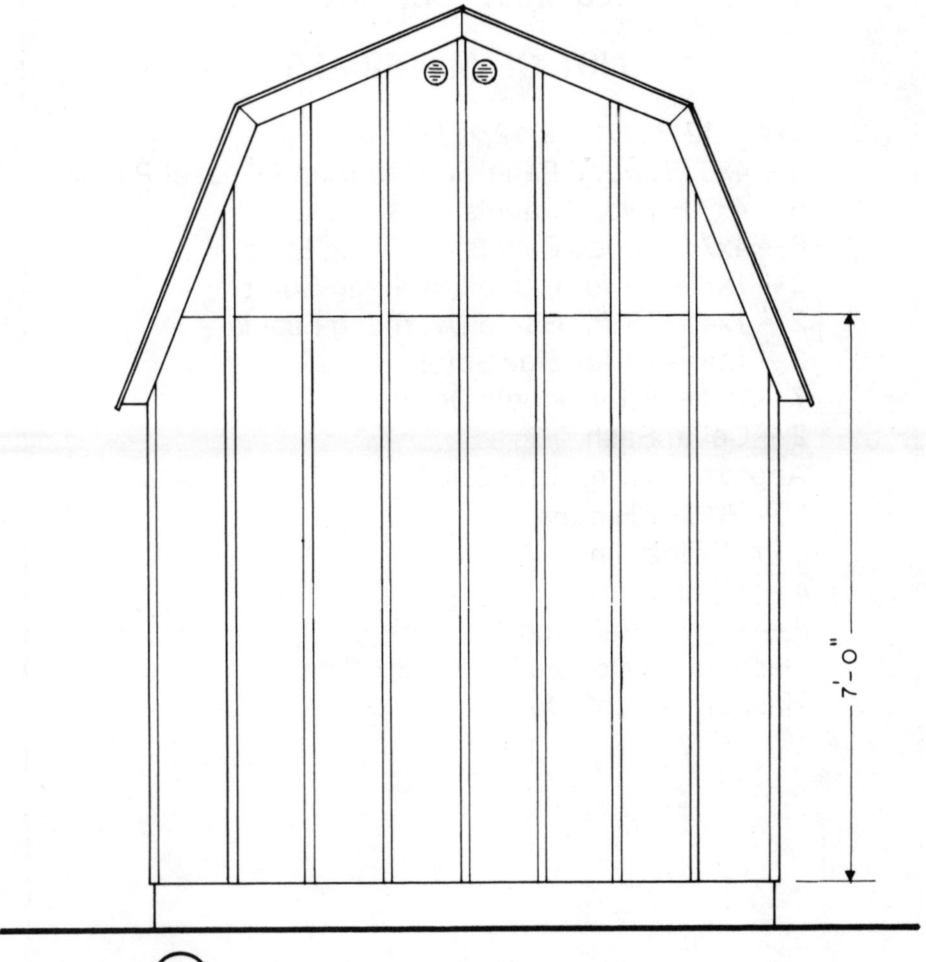

(197) **REAR ELEVATION**

RED BARN TOOL HOUSE

LIST OF MATERIALS

2x4 — 12/8, 4/10, 9/12, 13/14′
6 — 4x8 Plywood Panels for Roof and Gusset Plates
5 — 4x7 Plywood Panels for Sides
6 — 4x7 Plywood Panels for Gable Ends
2 — 1x6 — 16′ for Gable Overhang Soffit
2 — 1x4 — 16′ for Gable Overhang Fascia
2 — 1x4 — 12′ for Side Soffit
7 — 1x4 — 8′ for Double Door
2 — Cellar Sash
Approx. 300 lin. ft. 1x2 or ¾ x 1-5/8″ Battens
3 Pr. 6″ Tee Hinges
2 Pr. 4″ Tee Hinges
4 — 4″ Air Vents
8 — ½″x8″ Bolts and Washers
Flashing for Double Door — 5′x7″
Flashing for Loft Door — 4′x5″

HANDY - REFERENCE - LUMBER

PLYWOOD - FLAKEBOARD - HARDBOARD - MOLDINGS

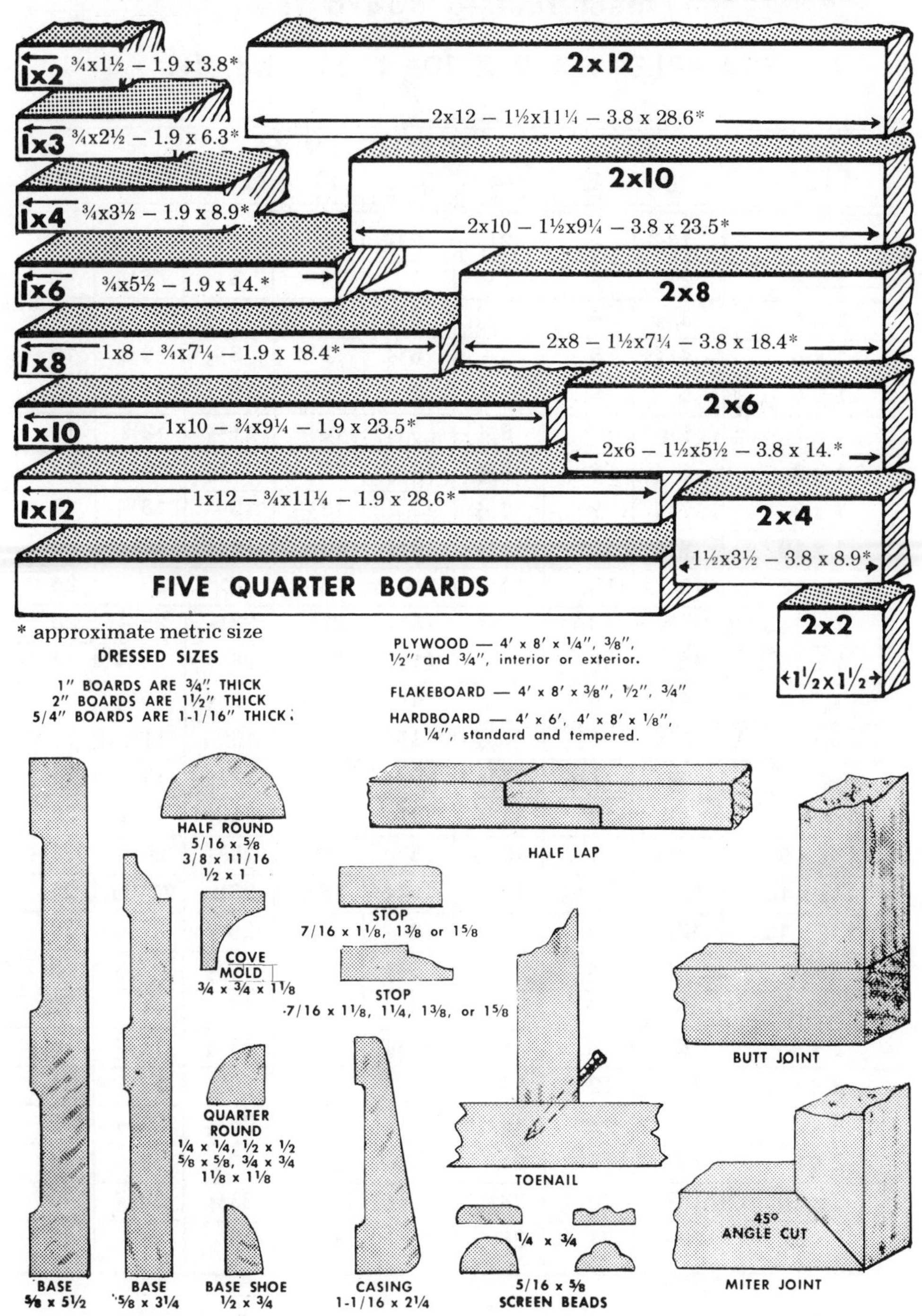
1x2 ¾x1½ – 1.9 x 3.8*
1x3 ¾x2½ – 1.9 x 6.3*
1x4 ¾x3½ – 1.9 x 8.9*
1x6 ¾x5½ – 1.9 x 14.*
1x8 1x8 – ¾x7¼ – 1.9 x 18.4*
1x10 1x10 – ¾x9¼ – 1.9 x 23.5*
1x12 1x12 – ¾x11¼ – 1.9 x 28.6*
FIVE QUARTER BOARDS
2x12
2x12 – 1½x11¼ – 3.8 x 28.6*
2x10
2x10 – 1½x9¼ – 3.8 x 23.5*
2x8
2x8 – 1½x7¼ – 3.8 x 18.4*
2x6
2x6 – 1½x5½ – 3.8 x 14.*
2x4
1½x3½ – 3.8 x 8.9*
2x2
1½x1½
* approximate metric size
DRESSED SIZES
1" BOARDS ARE ¾" THICK
2" BOARDS ARE 1½" THICK
5/4" BOARDS ARE 1-1/16" THICK
PLYWOOD — 4' x 8' x ¼", ⅜", ½" and ¾", interior or exterior.
FLAKEBOARD — 4' x 8' x ⅜", ½", ¾"
HARDBOARD — 4' x 6', 4' x 8' x ⅛", ¼", standard and tempered.
HALF ROUND
5/16 x ⅝
3/8 x 11/16
½ x 1
HALF LAP
STOP
7/16 x 1⅛, 1⅜ or 1⅝
COVE MOLD
¾ x ¾ x 1⅛
STOP
7/16 x 1⅛, 1¼, 1⅜, or 1⅝
BUTT JOINT
QUARTER ROUND
¼ x ¼, ½ x ½
⅝ x ⅝, ¾ x ¾
1⅛ x 1⅛
TOENAIL
¼ x ¾
45° ANGLE CUT
BASE
⅝ x 5½
BASE
⅝ x 3¼
BASE SHOE
½ x ¾
CASING
1-1/16 x 2¼
5/16 x ⅝
SCREEN BEADS
MITER JOINT

LUMBER CONVERSION CHART

Lineal feet to Board feet

EXAMPLE - 1 X 2 X 10 = 1-2/3 bd. ft.

SIZE	LENGTHS IN FEET 10	12	14	16	18	20	22	24
1 x 2	1⅔	2	2⅓	2⅔	3	3⅓	3⅔	4
1 x 3	2½	3	3½	4	4½	5	5½	6
1 x 4	3⅓	4	4⅔	5⅓	6	6⅔	7⅓	8
1 x 5	4⅙	5	5⅚	6⅔	7½	8⅓	9⅙	10
1 x 6	5	6	7	8	9	10	11	12
1 x 7	5⅚	7	8⅙	9⅓	10½	11⅔	12⅚	14
1 x 8	6⅔	8	9⅓	10⅔	12	13⅓	14⅔	16
1 x 9	7½	9	10½	12	13½	15	16½	18
1 x 10	8⅓	10	11⅔	13⅓	15	16⅔	18⅓	20
1 x 12	10	12	14	16	18	20	22	24
1 x 14	11⅔	14	16⅓	18⅔	21	23⅓	25⅔	28
1 x 16	13⅓	16	18⅔	21⅓	24	26⅔	29⅓	32
1¼ x 4	4⅙	5	5⅚	6⅔	7½	8⅓	9⅙	10
1¼ x 5	5⅙	6¼	7¼	8⅓	9⅛	10 5/12	11 5/12	12½
1¼ x 6	6¼	7½	8¾	10	11¼	12½	13¾	15
1¼ x 8	8⅓	10	11⅔	13⅓	15	16⅔	18⅓	20
1¼ x 9	9⅛	11¼	13 1/12	15	16⅚	18¾	20 7/12	22½
1¼ x 10	10 5/12	12½	14 7/12	16⅔	18¾	20⅚	22 11/12	25
1¼ x 12	12½	15	17½	20	22½	25	27½	30
2 x 2	3⅓	4	4⅔	5⅓	6	6⅔	7⅓	8
2 x 3	5	6	7	8	9	10	11	12
2 x 4	6⅔	8	9⅓	10⅔	12	13⅓	14⅔	16
2 x 6	10	12	14	16	18	20	22	24
2 x 8	13⅓	16	18⅔	21⅓	24	26⅔	29⅓	32
2 x 9	15	18	21	24	27	30	33	36
2 x 10	16⅔	20	23⅓	26⅔	30	33⅓	36⅔	40
2 x 12	20	24	28	32	36	40	44	48

HANDY REFERENCE-NAILS

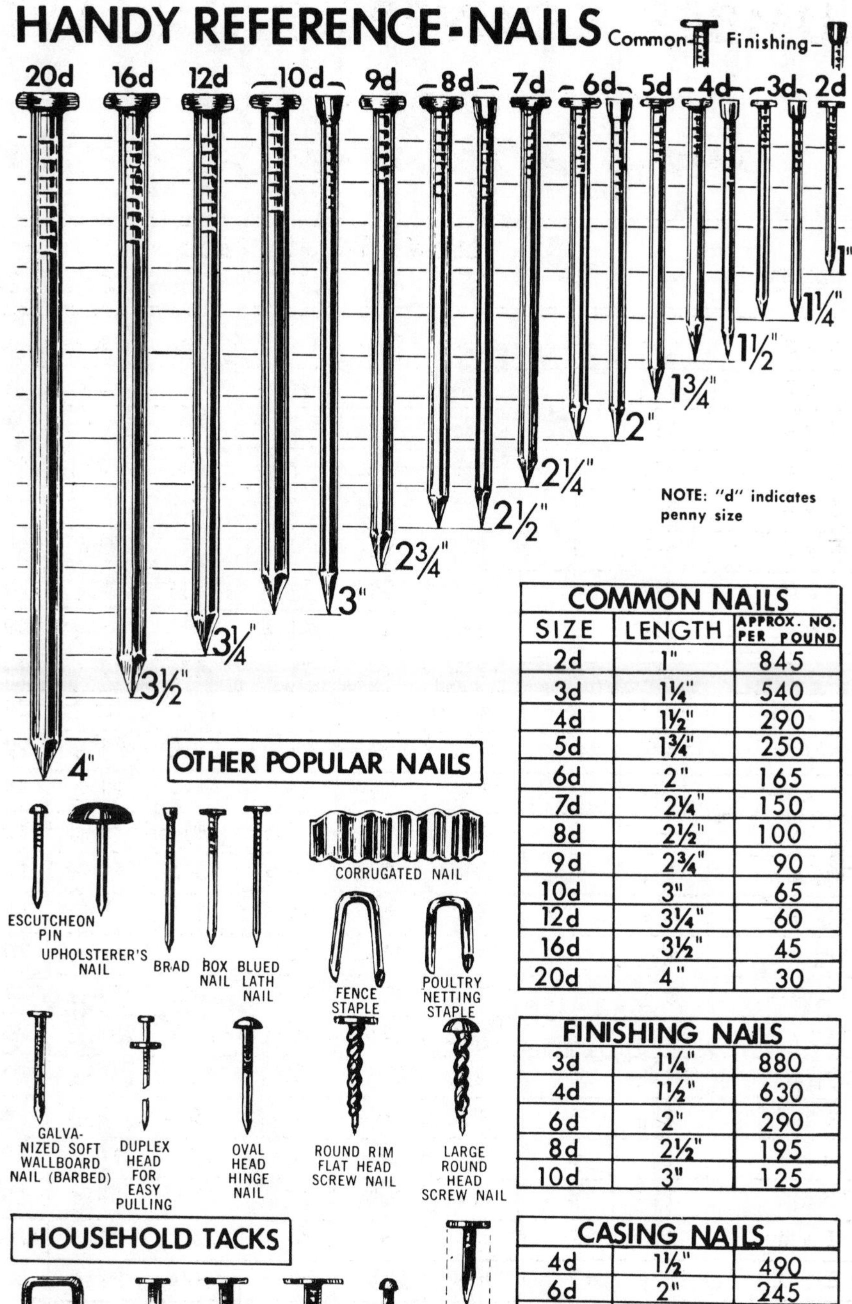

COMMON NAILS		
SIZE	LENGTH	APPROX. NO. PER POUND
2d	1"	845
3d	1¼"	540
4d	1½"	290
5d	1¾"	250
6d	2"	165
7d	2¼"	150
8d	2½"	100
9d	2¾"	90
10d	3"	65
12d	3¼"	60
16d	3½"	45
20d	4"	30

FINISHING NAILS		
3d	1¼"	880
4d	1½"	630
6d	2"	290
8d	2½"	195
10d	3"	125

CASING NAILS		
4d	1½"	490
6d	2"	245
8d	2½"	145
10d	3"	95
16d	3½"	72

HOUSEHOLD TACKS

DOUBLE POINTED TACK

UPHOLSTERER'S TACK

WIRE UPHOLSTERER'S TACK

BILL POSTER TACK

GIMP TACK

CHECKER HEAD CARPET TACK

HANDY REFERENCE-SCREWS

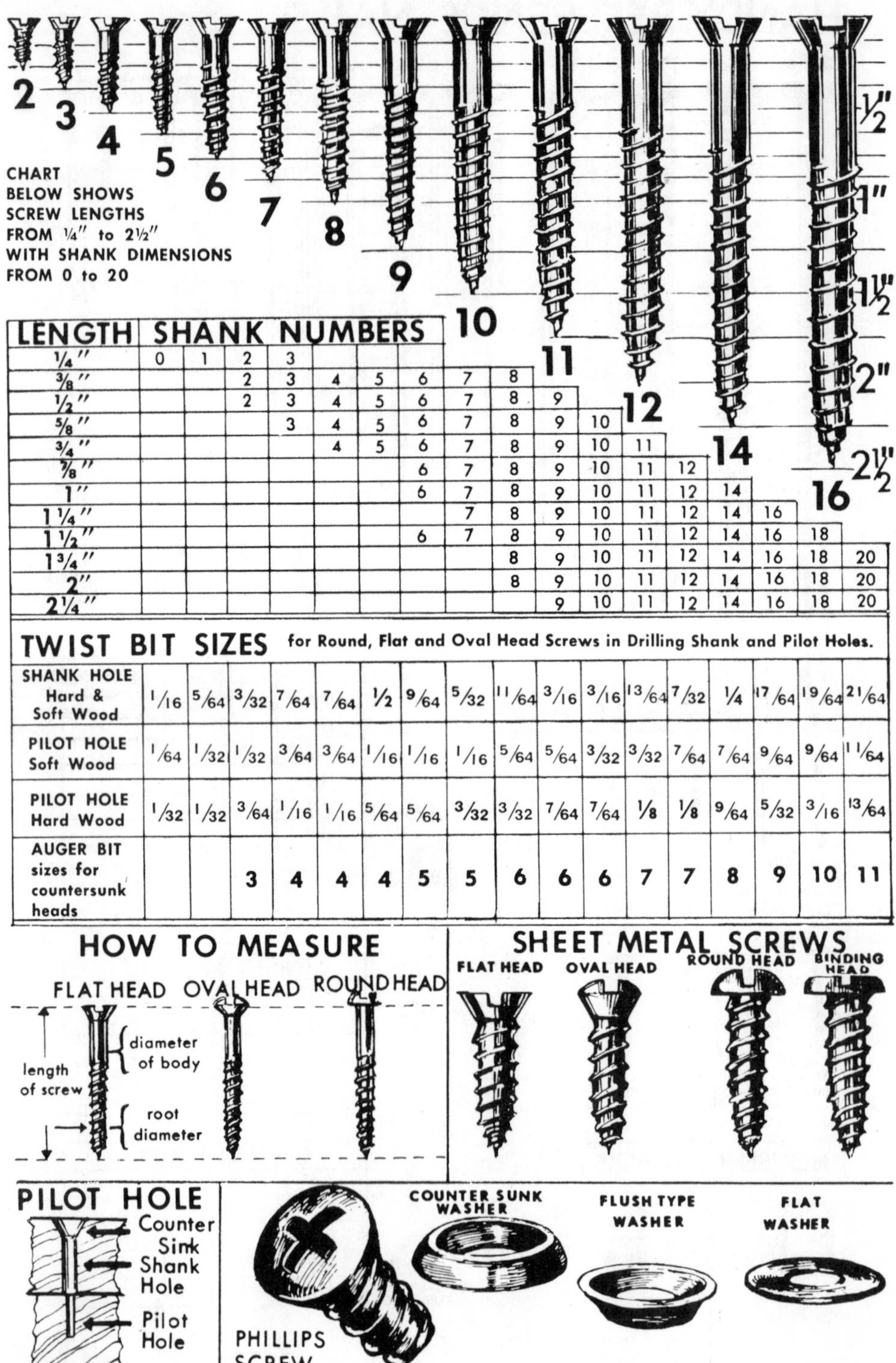

LENGTH	SHANK NUMBERS																
1/4"	0	1	2	3													
3/8"			2	3	4	5	6	7	8								
1/2"			2	3	4	5	6	7	8	9							
5/8"				3	4	5	6	7	8	9	10						
3/4"					4	5	6	7	8	9	10	11					
7/8"							6	7	8	9	10	11	12				
1"							6	7	8	9	10	11	12	14			
1 1/4"								7	8	9	10	11	12	14	16		
1 1/2"							6	7	8	9	10	11	12	14	16	18	
1 3/4"									8	9	10	11	12	14	16	18	20
2"									8	9	10	11	12	14	16	18	20
2 1/4"										9	10	11	12	14	16	18	20

TWIST BIT SIZES for Round, Flat and Oval Head Screws in Drilling Shank and Pilot Holes.

SHANK HOLE Hard & Soft Wood	1/16	5/64	3/32	7/64	7/64	1/2	9/64	5/32	11/64	3/16	3/16	13/64	7/32	1/4	17/64	19/64	21/64
PILOT HOLE Soft Wood	1/64	1/32	1/32	3/64	3/64	1/16	1/16	1/16	5/64	5/64	3/32	3/32	7/64	7/64	9/64	9/64	11/64
PILOT HOLE Hard Wood	1/32	1/32	3/64	1/16	1/16	5/64	5/64	3/32	3/32	7/64	7/64	1/8	1/8	9/64	5/32	3/16	13/64
AUGER BIT sizes for countersunk heads			3	4	4	4	5	5	6	6	6	7	7	8	9	10	11

The Nutcracker

For George with love,
Bernette

For my family,
Sam Williams

First North American edition published in 2012 by Boxer Books Limited.

www.boxerbooks.com

Library of Congress Cataloging-in-Publication Data available.

The illustrations were prepared using soft pencil and watercolour paint on hot press paper.
The text is set in Adobe Garamond

ISBN 978-1-907152-12-2

1 3 5 7 9 10 8 6 4 2

Printed in China

All of our papers are sourced from managed forests and renewable resources.

Ballet Kitty
Christmas Recital

Bernette Ford and Sam Williams

Boxer Books

Ballet Kitty woke up feeling happy!

Christmas was only days away, and Mademoiselle Felicity's Ballet School was putting on a big holiday recital.

Kitty wiggled her little pink toes into
her pink leotard and tights.

Then Kitty's best friend Princess Pussycat came over with a big picture book. It was the story of The Nutcracker.

The little dancers were not going to dance the whole ballet—just their favorite parts.

But they studied the book and listened to the music together.

"I can't wait for ballet class," said Kitty. "Mademoiselle will let us choose our parts today so we can get our costumes ready."

"I already know who I'm going to be," said Pussycat. "I'm going to dance the part of the Sugar Plum Fairy.

Her costume is lilac, my favorite color!"

"You can't be the Sugar Plum Fairy!" said Kitty.

"That's *my* favorite part.

And her costume is *not* lilac, it's pink!"

"It is not!" said Princess Pussycat. "It's lilac!"

Just then Ginger Tom arrived to walk

to ballet class with his friends.

"Why don't you two stop arguing?" said Tom.

"You're supposed to be best friends."

Kitty and Pussycat had no more to say.

They both knew Tom was right.

At ballet school, the little dancers gathered around Mademoiselle.

“Has everyone chosen a part?” she asked.

“I’m the Nutcracker!” said Tom.

"I'm the Sugar Plum Fairy!" said Kitty and Pussycat together.

"You can't both be the Sugar Plum Fairy," said Mademoiselle. "Wouldn't one of you like to be a Snowflake?"

That night in bed, a big tear rolled down Ballet Kitty's face. Princess Pussycat was her very best friend. She didn't want to be angry at her.

In her bed, Princess Pussycat sniffled. “I want to be the Sugar Plum Fairy,” she said. “But such a silly little thing should not keep me from talking to my best friend.”

A few days later, Ballet Kitty woke up happy.

She jumped out of bed and did a pirouette.

Kitty had had a great idea.

Princess Pussycat woke up feeling happy, too.
Tonight was the big Christmas recital,
and Pussycat had had a great idea.

That night at ballet school, Ginger Tom looked great as the Nutcracker.

When Princess Pussycat arrived, she was wearing a lovely Snowflake costume!

Then in came Ballet Kitty, dressed as
a beautiful Snowflake too!
The three little dancers jumped up and down
as they gave each other a big group hug.

Later, Kitty and Pussycat stood in the wings and watched Ginger Tom on stage. His dance was thrilling! Even Mademoiselle was smiling.

Then the girls floated onto the stage in their beautiful costumes. They heard the tinkling piano begin to play "The Waltz of the Snowflakes."

They pointed their toes.

They stretched
out their arms.

They bent and swayed
with the music.

The whole audience clapped and cheered.

The Christmas recital was a big success!

After the recital, the three friends walked arm in arm together. It had begun to snow.

They were going to Ballet Kitty's house for a holiday party. It was too soon for such a lovely night to end!